ENGINEERING LIBRARY

D0875859

Martin V. Butz

Rule-Based Evolutionary Online Learning Systems

Studies in Fuzziness and Soft Computing, Volume 191

Editor-in-chief
Prof. Janusz Kacprzyk
Systems Research Institute
Polish Academy of Sciences
ul. Newelska 6
01-447 Warsaw
Poland
E-mail: kacprzyk@ibspan.waw.pl

Further volumes of this series
can be found on our homepage:
springeronline.com

Vol. 175. Anna Maria Gil-Lafuente
Fuzzy Logic in Financial Analysis, 2005
ISBN 3-540-23213-3

Vol. 176. Udo Seiffert, Lakhmi C. Jain,
Patric Schweizer (Eds.)
*Bioinformatics Using Computational
Intelligence Paradigms,* 2005
ISBN 3-540-22901-9

Vol. 177. Lipo Wang (Ed.)
*Support Vector Machines: Theory and
Applications,* 2005
ISBN 3-540-24388-7

Vol. 178. Claude Ghaoui, Mitu Jain,
Vivek Bannore, Lakhmi C. Jain (Eds.)
Knowledge-Based Virtual Education, 2005
ISBN 3-540-25045-X

Vol. 179. Mircea Negoita,
Bernd Reusch (Eds.)
*Real World Applications of Computational
Intelligence,* 2005
ISBN 3-540-25006-9

Vol. 180. Wesley Chu,
Tsau Young Lin (Eds.)
Foundations and Advances in Data Mining,
2005
ISBN 3-540-25057-3

Vol. 181. Nadia Nedjah,
Luiza de Macedo Mourelle
Fuzzy Systems Engineering, 2005
ISBN 3-540-25322-X

Vol. 182. John N. Mordeson,
Kiran R. Bhutani, Azriel Rosenfeld
Fuzzy Group Theory, 2005
ISBN 3-540-25072-7

Vol. 183. Larry Bull, Tim Kovacs (Eds.)
Foundations of Learning Classifier Systems,
2005
ISBN 3-540-25073-5

Vol. 184. Barry G. Silverman, Ashlesha Jain,
Ajita Ichalkaranje, Lakhmi C. Jain (Eds.)
*Intelligent Paradigms for Healthcare
Enterprises,* 2005
ISBN 3-540-22903-5

Vol. 185. Spiros Sirmakessis (Ed.)
Knowledge Mining, 2005
ISBN 3-540-25070-0

Vol. 186. Radim Bělohlávek, Vilém
Vychodil
Fuzzy Equational Logic, 2005
ISBN 3-540-26254-7

Vol. 187. Zhong Li, Wolfgang A. Halang,
Guanrong Chen (Eds.)
*Integration of Fuzzy Logic and Chaos
Theory,* 2005
ISBN 3-540-26899-5

Vol. 188. James J. Buckley, Leonard J.
Jowers
Simulating Continuous Fuzzy Systems, 2006
ISBN 3-540-28455-9

Vol. 189. Hans-Walter Bandemer
Mathematics of Uncertainty, 2006
ISBN 3-540-28457-5

Vol. 190. Ying-ping Chen
*Extending the Scalability of Linkage
Learning Genetic Algorithms,* 2006
ISBN 3-540-28459-1

Vol. 191. Martin V. Butz
*Rule-Based Evolutionary Online Learning
Systems,* 2006
ISBN 3-540-25379-3

Martin V. Butz

Rule-Based Evolutionary Online Learning Systems

A Principled Approach to LCS Analysis and Design

 Springer

Dr. Martin V. Butz
Department of Cognitive Psychology
University of Würzburg
Röntgenring 11
97070, Würzburg
Germany
E-mail: mubtz@psychologie.uni-wuerzburz.de

Library of Congress Control Number: 2005932567

ISSN print edition: 1434-9922
ISSN electronic edition: 1860-0808
ISBN-10   3-540-25379-3 Springer Berlin Heidelberg New York
ISBN-13   978-3-540-25379-2 Springer Berlin Heidelberg New York

This work is subject to copyright. All rights are reserved, whether the whole or part of the material is concerned, specifically the rights of translation, reprinting, reuse of illustrations, recitation, broadcasting, reproduction on microfilm or in any other way, and storage in data banks. Duplication of this publication or parts thereof is permitted only under the provisions of the German Copyright Law of September 9, 1965, in its current version, and permission for use must always be obtained from Springer. Violations are liable for prosecution under the German Copyright Law.

Springer is a part of Springer Science+Business Media
springeronline.com
© Springer-Verlag Berlin Heidelberg 2006
Printed in The Netherlands

The use of general descriptive names, registered names, trademarks, etc. in this publication does not imply, even in the absence of a specific statement, that such names are exempt from the relevant protective laws and regulations and therefore free for general use.

Typesetting: by the author and TechBooks using a Springer LATEX macro package

Printed on acid-free paper   SPIN: 11370642   89/TechBooks   5 4 3 2 1 0

To my parents Susanne and Teja and my brother Christoph

Preface

Rule-based evolutionary online learning systems, often referred to as *Michigan-style learning classifier systems* (LCSs), were proposed nearly thirty years ago (Holland, 1976; Holland, 1977) originally calling them *cognitive systems*. LCSs combine the strength of reinforcement learning with the generalization capabilities of genetic algorithms promising a flexible, online generalizing, solely reinforcement dependent learning system. However, despite several initial successful applications of LCSs and their interesting relations with animal learning and cognition, understanding of the systems remained somewhat obscured. Questions concerning learning complexity or convergence remained unanswered. Performance in different problem types, problem structures, concept spaces, and hypothesis spaces stayed nearly unpredictable. This book has the following three major objectives: (1) to establish a facetwise theory approach for LCSs that promotes system analysis, understanding, and design; (2) to analyze, evaluate, and enhance the XCS classifier system (Wilson, 1995) by the means of the facetwise approach establishing a fundamental XCS learning theory; (3) to identify both the major advantages of an LCS-based learning approach as well as the most promising potential application areas. Achieving these three objectives leads to a rigorous understanding of LCS functioning that enables the successful application of LCSs to diverse problem types and problem domains. The quantitative analysis of XCS shows that the interactive, evolutionary-based online learning mechanism works machine learning competitively yielding a low-order polynomial learning complexity. Moreover, the facetwise analysis approach facilitates the successful design of more advanced LCSs including Holland's originally envisioned cognitive systems.

Martin V. Butz

Foreword I

In 1979, in an MIT library looking for inspiration, I encountered John Holland's (1978) book chapter, written with Judith Reitman, containing the first implementation of what we now call "learning classifier systems" (LCS). That led me to the stacks of MIT's Science Library and Holland's (1975) magnificent book *Adaptation in Natural and Artificial Systems* (which had not yet been taken out!). Devouring the first, and sampling as deeply as I could the second, I became hooked. How can a system improve in on-going interaction with its environment? How can it create new structures that permit better adaptation? How might programming just specify what to do, without saying how? Holland stated and aimed his work at exactly such questions, which, for understanding intelligence, I thought were exactly the right ones. From that point Holland's classifier systems became my main intellectual passion.

As is admirably related in Martin V. Butz's Introduction, the early path followed by the small band of LCS researchers had its ups and downs but, fortunately, persistence paid off and the field is now in rapid development, with over 700 published papers, several active research centers, and a prominent place in the major evolutionary computation conferences and journals.

Dr. Butz's book is in part a history and celebration of this progress, but much more importantly it is the first in-depth treatise available on learning classifier systems. He explains the relationships between LCS and its dual contexts of evolutionary computation and reinforcement learning . He discusses the basic structure of LCS systems and how David Goldberg's (2002) "facetwise" analysis can be applied to them. In the main part of the book, Butz presents in broad themes and solid detail the basic theory—much of which he originated—of the currently most-employed learning classifier system, XCS. At the same time, he provides a balanced representation of other LCS models and theories.

In later chapters Dr. Butz applies the theory—which is derived in a binary setting—to multiple-valued problems, data mining, and reinforcement learning. He discusses parameter and sub-system architecture selection. Finally, he presents his thinking about LCS extensions to systems with greater cognitive power that could include hierarchical structures and anticipatory behavior, among other properties.

I am truly excited by Martin Butz's great contribution and believe that this book will be turned to again and again for concepts, theory, and not least, examples of high-quality research.

Prediction Dynamics, Concord, MA   *Stewart W. Wilson*
wilson@prediction-dynamics.com

Foreword II

Toward the end of 1998, I started to correspond with a German graduate student named Martin Butz who wanted to visit my lab at Illinois. He was doing interesting work in so-called anticipatory classifier systems, and my early-1980s flirtation with learning classifier systems (LCSs) was ripe for resurrection. One thing led to another and Martin visited the Illinois Genetic Algorithms Laboratory, and I still remember our first meeting. He showed me some cool work, and I asked him some interesting questions, and the result was a spark, a flame, and a delightful collaboration that is still ongoing.

Martin's second book, Rule Based Evolutionary Online Learning Systems, is a (building?) blockbuster that should enrich the field, prolong the ongoing LCS renaissance, and inform the study and practice of genetics-based and other forms of machine learning. Monographs are like Isaiah Berlin's famous hedgehog: they are supposed to do one big thing, and many are lucky if they do that, but Butz has succeeded in doing two big things. First, he has published an effective design theory of learning classifier systems, and second, he has clearly demonstrated the value of linkage learning in an LCS context. The first of these things is important, because much of the theorizing about classifier systems has been qualitative and has avoided tough questions of complexity and convergence. Butz looks these twin antagonists in the eyes and with bounding calculations useful in both analysis and design. The linkage learning work clearly shows the importance of building-block processing in genetics-based machine learning in a manner that is unassailable, and in a way that appears to lead to regular solutions of extraordinarily difficult problems on a regular basis.

These two things would be enough, but like Babe Ruth pointing at the outfield, Butz finishes up with a call for more cognitive LCSs and a game plan for getting there. His plans are bold; they are almost brash, but I wouldn't bet against Butz, his energy, his perseverance, or his intelligence. Martin's book is a tale that is as yet unfinished, but I urge you to buy this book, read it, and stay tuned over the next ten years to see how it all turns out.

University of Illinois at Urbana-Champaign, IL   *David E. Goldberg*
deg@uiuc.edu

Acknowledgments

I am grateful to many people for supporting me not only intellectually but also mentally and socially in my work and life besides work. These acknowledgments can only give a glimpse on how much I benefited and learned from all my friends, family, and colleagues. Thank you so much to all of you.

I am in debt to my previous adviser David E. Goldberg, who supplied me with invaluable advice and guidance throughout my time at the University of Illinois at Urbana-Champaign (UIUC) concerning my research, writing, organization, and life. He supported me in all my plans and encouraged me to pursue thoughts and ideas that may have partially sounded very unconventional at first. Thank you so much for trusting and believing in me.

I am also very grateful to my other committee members during my PhD studies including Gerald DeJong, Sylvian Ray, and Dan Roth. Thank you for all the useful comments, suggestions, and additional work. I am also grateful to many other faculty members at UIUC including Gul Agha, Thomas Anastasio, Kathryn Bock, Gary Dell, Steven LaValle, and Edward Reingold. Thank you also for the support from the automated learning group at the national center for supercomputing applications (NCSA) and Michael Welge and Loretta Auvil, in particular.

In the mean time, I am very grateful to Stewart W. Wilson, who was always available for additional advice, support, and help for XCS and also my writing. My visit to Boston in April 2000 was more than inspiring and certainly shaped a large part of the basis of this book and the thoughts beyond it including the first complexity derivations for XCS.

I am also very grateful for all the support I received during my studies from Joachim Hoffmann, head of the Department of Cognitive Psychology at the Universität Würzburg. His cognitive perspective of things is always stimulating and many of the parts of this book related to cognition were shaped by the numerous discussions we had during my time in Germany.

I am also grateful to Wolfgang Stolzmann, who made my involvement into learning classifier systems and anticipatory systems possible in the first place and encouraged me to visit UIUC. Thank you for all the trust and encouragement.

I am also grateful to Pier Luca Lanzi, who grew to be more than a research partner and who showed me many aspects of XCS that I did not imagine in that way before—in particular in the reinforcement learning and function approximation side of things. His visit to UIUC was inspiring in many ways resulting in an additional research boost as well as an even closer lab community.

At the Illinois genetic algorithms laboratory (IlliGAL), I was blessed with a research environment of really smart people that were always available for advice and discussions but also knew when to talk about other issues in life. Particularly, I am very grateful to Kumara Sastry, student lab director and the best person to ask and discuss any problem that arises. I am similarly grateful to Xavier Llorà, who was a similar strong support, discussion partner, and inspiration besides the many other activities we shared. On my way to the PhD also Martin Pelikan was always an inspiration and a great help showing me new understandings and perspectives on many research issues and other concerns—unfortunately we could not meet as often anymore after Martin left IlliGAL in 2002. Thank you so much to all other current and former lab partners including Chang-Wook Ahn, Chen-Ju Chao, Ying-Ping Chen, Dimitri Knjazew, Fiepeng Li, Fernando Lobo, Jonathan Loveall, Naohiro Matsumura, Abhishek Sinha, Kurian Tharakunnel, Tian-Li Yu and the lab visitors Hussein Abbass, Jaume Bacardit, Clarissa Van Hoyweghen, Pier Luca Lanzi, Claudio Lima, Kei Onishi, Gerulf Pedersen, and Franz Rothlauf. The close cooperations with Hussein Abbass, Jaume Bacardit and Pier Luca Lanzi were particularly fruitful. Thank you all for being so good friends and taking care of the lab and other things together.

The book significantly improved due to comments from many people including David Goldberg, Oliver Herbort, Marjorie Kinney, Pier Luca Lanzi, Martin Pelikan, Kumara Sastry, Samarth Swarup, and Stewart Wilson. Richmond Kinney carefully proofread the book once more in its final form. Thank you for all the help.

Thank you to all my other friends and research mates at UIUC and beyond including Jason Bant, Jacob Borgerson, Amanda Hinkle, John Horstman, Alex Kosorukov, Kiran Lakkaraju, Samarth Swarup, Dav Zimak and many others. I was also blessed with and greatly supported by my girlfriend Marjorie Kinney, who always listens to my thoughts and is an inspiration in many ways—thank you!

I am also grateful to my colleagues and friends back in Germany. The team at the department of cognitive psychology was a great help and made me feel always welcome and happy when I came back. Thank you in particular to Andrea Kiesel. Andrea was always helpful and more than simply supportive she made me feel home—thank you! Thank you for all the cooperation and help from all the others in the department including Andrea Albert, Michael Berner, Oliver Herbort, Silvia Jahn, Wilfried Kunde, Alexandra Lenhard,

Georg Schüssler, Albrecht Sebald, Armin Stock, and Annika Wagener. I am looking forward to the next few years!

Thank you to all my other friends back home that were always available and always welcoming me including Marianne Greubel, Heiko and Frank Hofmann, Gregor Kopka, Stefan Merz, Rainer Munzert, André Ponsong, Matthias Reiher, Franz Rothlauf, Peter Scheerer, Marion Schmidt, Daniel Schneider and Johannes Schunk. I don't know what I would do without you.

I am also very grateful to my whole family back in Germany. My brother Christoph is an inspiration on his own. My parents, Susanne and Teja, helped me keep the balance between the US and Germany and supported me in various ways. I am glad we grew so close together again despite these long distances for such long times.

Finally, I would like to acknowledge all the financial support I received over the last years including money from the German research foundation (Deutsche Forschungsgemeinschaft, DFG) under grant DFG HO1301/4-3, the European commission contract no. FP6-511931, the National Center for Supercomputing Applications (NCSA) at UIUC, as well as the CSE fellowship I received from the Computational Science and Engineering department at UIUC during my PhD studies. Thank you also for the additional money from other resources available to the IlliGAL lab at UIUC as well as to the Department of Cognitive Psychology at the Universität Würzburg that supplied me with computer power and an optimal working environment.

Contents

1 Introduction 1

2 Prerequisites 9
    2.1 Problem Types 10
        2.1.1 Optimization Problems 10
        2.1.2 Classification Problems 13
        2.1.3 Reinforcement Learning Problems 15
    2.2 Reinforcement Learning 18
        2.2.1 Model-Free Reinforcement Learning 18
        2.2.2 Model-Based Reinforcement Learning 21
    2.3 Genetic Algorithms 21
        2.3.1 Basic Genetic Algorithm 22
        2.3.2 Facetwise GA Theory 22
        2.3.3 Selection and Replacement Techniques 23
        2.3.4 Adding Mutation and Recombination 26
        2.3.5 Component Interaction 28
    2.4 Summary and Conclusions 28

3 Simple Learning Classifier Systems 31
    3.1 Learning Architecture 32
        3.1.1 Knowledge Representation 32
        3.1.2 Reinforcement Learning Component 34
        3.1.3 Evolutionary Component 35
    3.2 Simple LCS at Work 36
    3.3 Towards a Facetwise LCS Theory 37
        3.3.1 Problem Solutions and Fitness Guidance 38
        3.3.2 General Problem Solutions 39
        3.3.3 Growth of Promising Subsolutions 40
        3.3.4 Neighborhood Search 41
        3.3.5 Solution Sustenance 47
        3.3.6 Additional Multistep Challenges 47

        3.3.7   Facetwise LCS Theory 48
    3.4   Summary and Conclusions 49

4   **The XCS Classifier System** 51
    4.1   System Introduction 51
        4.1.1   Knowledge Representation 52
        4.1.2   Learning Interaction 52
        4.1.3   Rule Evaluation 53
        4.1.4   Rule Evolution 56
        4.1.5   XCS Learning Intuition 57
    4.2   Simple XCS Applications 58
        4.2.1   Simple Classification Problem 58
        4.2.2   Performance in Maze 1 61
    4.3   Summary and Conclusions 63

5   **How XCS Works: Ensuring Effective Evolutionary**
    Pressures 65
    5.1   Evolutionary Pressures in XCS 66
        5.1.1   Semantic Generalization due to Set Pressure 67
        5.1.2   Mutation's Influence 69
        5.1.3   Deletion Pressure 70
        5.1.4   Subsumption Pressure 70
        5.1.5   Fitness Pressure 71
        5.1.6   Pressure Interaction 72
        5.1.7   Validation of the Specificity Equation 73
    5.2   Improving Fitness Pressure 80
        5.2.1   Proportionate vs. Tournament Selection 81
        5.2.2   Limitations of Proportionate Selection 82
        5.2.3   Tournament Selection 84
    5.3   Summary and Conclusions 88

6   **When XCS Works: Towards Computational Complexity** ... 91
    6.1   Proper Population Initialization:
          The Covering Bound 92
    6.2   Ensuring Supply: The Schema Bound 94
        6.2.1   Population Size Bound 95
        6.2.2   Specificity Bound 96
        6.2.3   Extension in Time 97
    6.3   Making Time for Growth: The Reproductive Opportunity Bound 98
        6.3.1   General Population Size Bound 98
        6.3.2   General Reproductive Opportunity Bound 100
        6.3.3   Sufficiently Accurate Values 102
        6.3.4   Bound Verification 103
    6.4   Estimating Learning Time 103
        6.4.1   Time Bound Derivation 104

        6.4.2   Experimental Validation 106
    6.5 Assuring Solution Sustenance: The Niche Support Bound 108
        6.5.1   Markov Chain Model 109
        6.5.2   Steady State Derivation 111
        6.5.3   Evaluation of Niche Support Distribution 112
        6.5.4   Population Size Bound 116
    6.6 Towards PAC Learnability 117
        6.6.1   Problem Bounds Revisited 118
        6.6.2   PAC-Learning with XCS 120
    6.7 Summary and Conclusions 121

7   Effective XCS Search: Building Block Processing 123
    7.1 Building Block Hard Problems 124
        7.1.1   Fitness Guidance Exhibited 125
        7.1.2   Building Blocks in Classification Problems 126
        7.1.3   The Need for Effective BB Processing 128
    7.2 Building Block Identification and Processing 130
        7.2.1   Structure Identification Mechanisms 132
        7.2.2   BB-Identification in the ECGA 132
        7.2.3   BB-Identification in BOA 134
        7.2.4   Learning Dependency Structures in XCS 137
        7.2.5   Sampling from the Learned Dependency Structures 138
        7.2.6   Experimental Evaluation 141
    7.3 Summary and Conclusions 144

8   XCS in Binary Classification Problems 147
    8.1 Multiplexer Problem Analyses 147
        8.1.1   Large Multiplexer Problems 148
        8.1.2   Very Noisy Problems 150
    8.2 The xy-Biased Multiplexer 151
    8.3 Count Ones Problems 152
    8.4 Summary and Conclusions 154

9   XCS in Multi-Valued Problems 157
    9.1 XCS with Hyperrectangular Conditions 158
    9.2 Theory Modifications 159
        9.2.1   Evolutionary Pressure Adjustment 159
        9.2.2   Population Initialization: Covering Bound 160
        9.2.3   Schema Supply and Growth: Schema
                and Reproductive Opportunity Bound 160
        9.2.4   Solution Sustenance: Niche Frequencies 162
        9.2.5   Obliqueness 162
    9.3 Datamining 163
        9.3.1   Datasets 163
        9.3.2   Results 165

9.4   Function Approximation 169
      9.4.1 Hyperellipsoidal Conditions 169
      9.4.2 Performance Comparisons 172
9.5   Summary and Conclusions 178

10 XCS in Reinforcement Learning Problems 181
10.1 Q-learning and Generalization 182
10.2 Ensuring Accurate Reward Propagation:
      XCS with Gradient Descent 184
      10.2.1 Q-Value Estimations and Update Mechanisms 184
      10.2.2 Adding Gradient Descent to XCS 185
10.3 Experimental Validation 186
      10.3.1 Multistep Maze Problems 187
      10.3.2 Blocks-World Problems 192
10.4 Summary and Conclusions 194

11 Facetwise LCS Design 197
11.1 Which Fitness for Which Solution? 197
11.2 Parameter Estimation 198
11.3 Generalization Mechanisms 200
11.4 Which Selection For Solution Growth? 202
11.5 Niching for Solution Sustenance 203
11.6 Effective Evolutionary Search 204
11.7 Different Condition Structures 205
11.8 Summary and Conclusions 206

12 Towards Cognitive Learning Classifier Systems 207
12.1 Predictive and Associative Representations 208
12.2 Incremental Learning 210
12.3 Hierarchical Architectures 212
12.4 Anticipatory Behavior 214
12.5 Summary and Conclusions 216

13 Summary and Conclusions 219
13.1 Summary 219
13.2 Conclusions 223

A Notation and Parameters 227

B Algorithmic Description 231

C Boolean Function Problems 243
      Multiplexer 243
      Layered Multiplexer 244
      xy-Biased Multiplexer 244
      Hidden Parity 245

Count Ones 246
Layered Count Ones 246
Carry Problem 247
Hierarchically Composed Problems 248

References 249

Index 263

1

Introduction

Rule-based evolutionary online learning systems, often referred to as *Michigan-style learning classifier systems* (LCSs) [1], were originally inspired by the general principles of Darwinian evolution and cognitive learning. In fact, when John Holland proposed the basic LCS framework (Holland, 1976; Holland, 1977; Holland & Reitman, 1978), he actually referred to LCSs as *cognitive systems* (CSs). Inspired by stimulus-response principles in cognitive psychology, Holland designed CSs to evolve a set of production rules that convert given input into useful output. Temporary memory in the form of a *message list* was added to simulate inner mental states situating the system in the current environmental context.

Early work on LCSs confirmed the great potential of the systems for simulating animal learning and cognition as well as for real-world applications. In the first classifier system implementation, Holland and Reitman (1978) confirmed that LCSs can simulate animal behavior successfully. They evolved a representation that resulted in goal-directed, stimulus-response-based behavior satisfying multiple goals represented in resource reservoirs. Booker (1982) extended Holland's approach by experimenting with an agent that needs to avoid aversive stimuli and reach attractive stimuli. Wilson (Wilson, 1985; Wilson, 1987a) confirmed the potential of LCSs to simulate artificial animals termed *animats*—triggering the animat approach to artificial intelligence (Wilson, 1991). In brief, the approach suggests simulating animats in simulated, progressively more complex environments to understand

---

[1] This book is concerned with Michigan-style LCSs. These systems are online learning systems, which iteratively interact with a problem, receiving one problem instance at a time. In contrast to Michigan-style LCSs, there are Pittsburgh-style learning classifier systems (DeJong, Spears, & Gordon, 1993; Llorà & Garrell, 2001b; Llorà & Garrell, 2001a), which are batch learning systems that are much more similar to pure genetic algorithms. Despite the fundamental differences of batch-learning and the evolution of a set of solutions in these systems, a big part of the analysis in this book may be carried over—appropriately modified—to Pittsburgh-style LCSs.

learning in organisms as well as to develop highly adaptive autonomous robotic systems. Goldberg (1983) successfully applied an LCS to the control of a simulated pipeline system, confirming that LCSs are valuable learning systems for real-world applications as well.

Many of these publications were far-reaching and somewhat visionary. The LCS framework predated and inspired the now well-established reinforcement learning field (Kaelbling, Littman, & Moore, 1996; Sutton & Barto, 1998). The originally used *bucket-brigade* algorithm in LCSs (Holland, 1985) distributed reward very similar to now well-established temporal difference learning techniques, such as TD(λ) or SARSA (Sutton & Barto, 1998). The ambitious scenarios and the relation to animal learning, cognition, and robotics pointed towards research directions that remain mind boggling even today. Thus, most early LCS work laid out very interesting and challenging future research directions.

Despite these promising factors and interesting directions, the LCS framework was somewhat ahead of its time. Due to the high complexity of the systems, scalable learning of robust problem solutions could not be guaranteed. Essentially, hardly any theory was developed for an LCS system, because (1) neither learning nor convergence could be assured mathematically, (2) the learning interactions in the system appeared to be too complex and remained not well-understood, (3) the learning biases of the system were only explained intuitively, and (4) competitive applications were restricted to a somewhat limited set of smaller problems. These problematic factors led a surprisingly wide inacceptance of LCSs in the artificial intelligence and machine learning literature.

In their *"Critical review of classifier systems"*, Wilson and Goldberg (1989) pointed out several of the most important problems in the available LCSs at that time. First, it appeared that successful reward chains were hard to learn and to maintain by the means of the *bucket-brigade* algorithm in combination with the evolutionary component. Second, inappropriate bidding and payment schemes obstructed generalization, enabled overgeneralization, or prevented the formation of default hierarchies. Third, the limitations of simple classifier syntax prohibited effective processing of noisy input features, continuous problem spaces, or larger binary problem spaces. Besides these challenges, Wilson and Goldberg (1989) also mentioned the importance of developing and understanding planning and lookahead mechanisms, representations for expectations, implementations of a short-term memory, and population sizing equations.

During the subsequent LCS winter, Stewart Wilson and a few others continued to work in the LCS field. And it was Stewart Wilson who heralded an LCS renaissance with the publication of the two most influential LCS systems to date: (1) the zeroth level classifier system ZCS (Wilson, 1994) and (2) the accuracy-based classifier system XCS (Wilson, 1995).

Both classifier systems overcome many of the previously encountered challenges. The credit assignment mechanism in ZCS and XCS is directly related to the then well-understood Q-learning algorithm (Watkins, 1989) in the reinforcement learning (RL) literature so that appropriate reward estimation and propagation is ensured. Overgeneralization problems are overcome by proper fitness sharing techniques (ZCS) or the new accuracy-based fitness approach (XCS). Additionally, in XCS generalization is achieved by a niche reproduction combined with population-wide deletion, as stated in Wilson's generalization hypothesis (Wilson, 1995).

Published results suggested the competitiveness of the new LCSs (Wilson, 1994; Wilson, 1995). Solutions were found in previously unsolved maze problems that require proper generalization as well as hard Boolean function problems, such as the multiplexer problem.

Later, research focused further on the XCS system solving larger Boolean function problems (Wilson, 1998), suggesting the scalability of the system. Others focused on performance investigations in larger maze problems considering action noise and generalization (Lanzi, 1997; Lanzi, 1999a; Lanzi, 1999c).

In addition to the promising experimental results, the growth of qualitative and quantitative theoretical insights and understanding slowly gained momentum. Tim Kovacs investigated Wilson's generality hypothesis in more detail and showed that XCS strives to learn complete, accurate, and minimal representations of Boolean function problems (Kovacs, 1997). Later, Kovacs investigated the appropriate fitness approach in LCSs, contrasting a purely strength-based approach with XCS's accuracy-based approach (Kovacs, 2000; Kovacs, 2001). Finally, one of the most important questions was asked: *What makes a problem hard for XCS* (Kovacs & Kerber, 2001)? This question led to some insights on problem difficulty with respect to the optimal solution representation [O]. However, it remained obscured *how* XCS evolves an optimal solution as well as *which computational requirements* are necessary to successfully evolve and maintain such a solution.

Besides the new direct insights into LCSs, genetic algorithms (GAs) are now much better understood than they were back in the late 1980s. Goldberg provided a comprehensive introduction to GAs (Goldberg, 1989) and later suggested a *facetwise approach* to GA theory and design (Goldberg, 1991). The facetwise approach puts forward a modular analysis of GA components and their interactions including different selection types, structure propagation and disruption via recombination, mutation influences, or structure sustenance. The approach enabled a rigorous quantitative analysis of GA components and their interaction, which lead to a proper understanding of GA scale-up behavior and its dependence on problem structure (Goldberg, Deb, & Clark, 1992; Goldberg, Deb, & Thierens, 1993; Harik, Cantú-Paz, Goldberg, & Miller, 1997; Goldberg, 1999; Goldberg, 2002).

In addition to the *quantitative* achievements, the design decomposition also enables a rigorous *qualitative* understanding of what GAs are really searching for. Holland (1975) already hypothesized that GAs are processing *schemata*, referring to low-order attribute dependencies. GA learning success depends on the successful detection and propagation of such dependencies. However, Holland's original schema theory mainly showed the potential failure of schema processing instead of focusing on the best way to identify and propagate *useful schemata*, which are often called *building blocks* (BBs) (Holland, 1975; Goldberg, 1989; Goldberg, 2002). BBs may be characterized as lower order dependency structures (in which few attributes interact) that result in a fitness increase when set to the correct values. Attributes of a BB structure interact nonlinearly with respect to their fitness influence so that a small difference from the correct values may lead to a large difference in fitness. It should be clear that the presence of BBs highly depends on the problem at hand as well as on the chosen problem representation (Rothlauf, 2002).

It should be noted that Goldberg's facetwise analysis approach does not only facilitate system analysis and modeling but also leads to a more general system understanding and enables more effective system design (Goldberg, 2002). In the pure GA realm, for example, the GA design decomposition led to the creation of *competent GAs*—GAs that solve boundedly difficult problems quickly, accurately and reliably—including the extended compact GA (ECGA) (Harik, 1999) and the Bayesian optimization algorithm (BOA) (Pelikan, Goldberg, & Cantu-Paz, 1999) and triggering the field of estimation of distribution algorithms (EDAs) (Mühlenbein & Paaß, 1996; Pelikan, Goldberg, & Lobo, 2002; Larrañaga, 2002).

Objectives

The train of thought in this book follows a similar decomposition approach to analyze LCSs. With Wilson's powerful XCS system at hand, it establishes a rigorous understanding of XCS functioning, computational requirements, convergence properties, and generalization capabilities. The design decomposition enables us to consider evolutionary components independently so that a precise and general system analysis is possible. Meanwhile, the analysis leads us to several successfully integrated system improvements. Moreover, the proposed decomposition points towards many interesting prospective research directions including further LCS analyses and the modular and hierarchical design of more advanced LCSs.

In further detail, we first establish a rigorous understanding of LCSs, and the XCS classifier system in particular. We show which learning mechanisms can be identified, which learning biases these mechanisms cause, and how they interact. The undertaken facetwise analysis enables us to establish a fundamental theory for population sizing, problem difficulty, and learning speed. It is shown that the derived problem bounds can be used to confirm (restricted) PAC-learning capabilities of the XCS system in k-*DNF* problems. Moreover, the analysis leads us to the identification and analysis of BB-hard problems in the LCS realm. We consequently integrate competent GA recombination operators solving the BB-hard problems by making evolutionary search more effective.

Additionally, we draw connections with neural network-based function approximation techniques, combined with reinforcement learning mechanisms (Baird, 1999; Haykin, 1999; Sutton & Barto, 1998) and tabular Q-learning (Watkins, 1989). It is shown that the integration of gradient techniques improves learning reliability and accuracy in XCS. The theoretical considerations confirm that LCSs are hybrid techniques that have neural network-like interdependence properties but also tabular-like independence properties.

Besides the theoretical and mechanism-based enhancements, the book shows key-results in various problem domains including binary, nominal, and real-valued classification problems as well as multistep RL problems. Learning behavior is analyzed respecting typical problem structures and problem properties.

The lessons learned from the XCS analysis provide a broader understanding of LCSs and their interactive learning mechanisms. The analyses and experimental evaluations combined with the facetwise approach lay out a clear path towards the successful design and application of future LCS-based learning systems. The key to success is an appropriate combination of necessary learning biases comprised in the structure and type of LCS modules and their efficient interaction. Representational considerations are thereby as important as the choice of mechanisms and their effective integration.

With the gained understanding at hand, we finally propose the creation of LCS-based, cognitive learning systems that may learn interactively and incrementally a modular, distributed, and hierarchical predictive problem representation and use the representation to pursue anticipatory cognitive behavior. The proposed cognitive LCS-based structures are in accordance with Holland's original ideas but are now endowed with a modular theory on computational requirements, interactivity, learning reliability, solution accuracy and quality as the supporting backbone. The book lies out the foundations for the successful creation and application of such competent modular integrative LCS-based learning structures.

Road Map

The remainder of the book is structured as follows:

Chapter 2 provides an overview of required background knowledge. First, we introduce optimization, classification and RL problems and discuss most important structural properties, differences, and problem difficulties. Next, we provide an overview of relevant RL mechanisms. Finally, we introduce GAs focusing on Goldberg's facetwise GA decomposition approach and the aspects within most relevant for our subsequent analyses on LCSs.

Chapter 3 first gives a gentle introduction to a basic LCS system. The application to a simple toy problem illustrates the general functioning. Next, we discuss LCS theory and analysis. In particular, we propose a facetwise LCS theory approach decomposing the LCS architecture into relatively independent system facets. Each facet needs to be implemented appropriately to ensure the successful system application.

Chapter 4 introduces the system under investigation, that is, the accuracy-based classifier system XCS (Wilson, 1995). We illustrate XCS's learning behavior on exemplar toy problems, including classification and RL problems, revealing basic intuition behind XCS functioning. We then proceed to our XCS analysis.

XCS's major learning biases are investigated in Chapter 5. We show that fitness propagates accurate rules whereas generalization is achieved by a combination of subset-based reproduction and population-wide deletion. We derive a specificity equation that models the behavior of specificity in a population not influenced by fitness. Finally, we replace the previously applied proportionate selection with a subset-size dependent tournament selection mechanism ensuring reliable fitness pressure towards better classifiers.

Chapter 6 analyzes the computational requirements for solution growth and sustenance. We show that initial specificity and population size needs to be chosen adequately to ensure learning startup, minimal structural supply, relevant structural growth, and solution sustenance. With the additional learning time estimate, we can show that the computational effort scales in a low-order polynomial in problem length and solution complexity.

Next, we address solution search. Chapter 7 confirms that effective BB structure identification and processing may be necessary also in the realms of classification and RL. We introduce statistical techniques to extract evolved lower level problem structure. The gained knowledge about dependency structures is then used to mutate and recombine offspring rules more effectively, consequently solving previously hard problems successfully.

Chapter 8 applies the resulting XCS system to diverse Boolean function problems. We investigate performance in large problems, the impact of irrelevant problem features, overlapping problem subsolutions, unequally distributed subsolution complexities, and external noise. As a whole, the chapter experimentally confirms the theoretic learning bounds and supports the derived mathematical learning robustness and scalability results.

Chapter 9 applies the XCS system to real-world datamining problems as well as function approximation problems. We compare XCS's performance with several other machine learning systems in the investigated datasets. The comparison further confirms XCS's learning competence and machine learning competitiveness. We also enhance the facetwise theory to the real-valued problem domain.

Chapter 10 then investigates multistep RL problems addressing the additional challenges of reward backpropagation and distribution. The chapter shows that XCS is a competent online generalizing RL system that is able to ignore additional irrelevant problem features with additional computational effort that is linear in the number of features. The results confirm that XCS offers a robust alternative to purely neural-based RL approaches.

With pieces of the LCS puzzle then in place, Chapter 11 outlines how a similar facetwise problem approach may be helpful in the analysis of other LCSs and how LCSs can be designed using the modular theory approach. Alternative learning biases are discussed as well.

Chapter 12 then outlines how the analysis may carry over to the design of further competent and flexible LCS systems targeted to the problem at hand. In particular, we put forward the integration of LCS learning mechanisms into cognitive learning structures. Hierarchical and modular structures, anticipatory mechanisms, incremental learning, and sequential processing mechanisms are discussed. LCS mechanisms may serve as the tool-box for generating the desired structures.

Chapter 13 summarizes the major findings. The conclusions outline the next steps necessary for further LCS analysis and more competent LCS design. With the facetwise perspective on LCSs at hand, the design of Holland's originally envisioned cognitive systems may finally be within our grasp.

2

Prerequisites

LCSs are designed to solve classification as well as more general *reinforcement learning* (RL) problems. LCSs solve these problems by evolving a rule-base of classifiers by the means of a critic based on RL techniques for rule evaluation and a GA for rule evolution. Before jumping directly into the LCS arena, we first look at these prerequisites.

Since optimization and learning is comparable to a *search for expected structure*, we first look at the problem types and problem structures we are interested in. We differentiate between *optimization problems*, *classification problems*, and *Markov decision process problems*. Each problem causes different but related challenges. Thus, successful learning architectures need to be endowed with different but related learning mechanisms and learning biases. LCSs are facing RL problems but might also be applied to classification or even optimization problems.

Since it is provenly impossible that there exists a learning technique that can solve all possible problems (assuming that all expressible problems given a certain representation are equally likely) better than simple enumeration or random search (Wolpert & Macready, 1995; Wolpert, 1996b; Wolpert, 1996a; Wolpert & Macready, 1997), it is important to characterize and distinguish different problem structures. In this problem introduction we focus on such different problem properties. We then use these properties to reveal what GAs and LCSs are actually searching for; in other words, we identify the learning biases of the algorithms. As we will see, GAs and LCSs are very general learning mechanisms that are searching for lower order dependency structures in a problem, often referred to as building blocks (BBs). The advantage of such a search bias is that natural problems typically but arguably obey such a hierarchically composed problem structure (Simon, 1969; Gibson, 1979; Goldberg, 2002).

Apart from the necessary understanding of LCS-relevant problem types and typical problem structures, the second major prerequisite is a general understanding of RL techniques and *genetic algorithms* (GAs). Section 2.2 introduces RL including the most relevant Q-learning algorithm (Watkins,

1989). We show that RL is well-suited to serve as an online learning actor/critic system that is capable of evaluating rules, distributing reward, and making action (or classification) decisions. Section 2.3 introduces GAs, which are well-suited to learn relevant rule structures given a quality or *fitness* measure. Additionally, we highlight the importance of the mentioned *facetwise analysis* approach taken in the GA literature to promote understanding of GA functioning, scale-up behavior, and parameter settings as well as to enable the design of more elaborate, *competent GAs* (Goldberg, 2002).

Summary and conclusions summarize the most important issues addressed in this chapter, pointing towards the integration of the addressed mechanisms into LCSs.

2.1 Problem Types

LCSs may be applied in two major problem domains. One is the world of classification problems. The second is the one of RL problems either modeled by a Markov decision process (MDP) or by a more general, partially observable Markov decision process (POMDP).

In classification problems, feedback is provided instantly. Successive problem instances are usually sampled independently and identically distributed. On the other hand, in RL problems feedback may be delayed in time and successive problem input depends on the underlying problem structure as well as on the chosen actions. Thus, internal reinforcement propagation becomes necessary, which poses an additional learning challenge.

Before introducing classification and RL problems, we give a short introduction to optimization problems emphasizing similarities and differences as well as typically expectable problem structures and properties.

2.1.1 Optimization Problems

An optimization problem is a problem in which a particular structure or solution needs to be optimized. Thus, given a particular solution space, an optimization algorithm searches for the best solution in the solution space. Optimization problems cover a huge range of problems such as (1) the optimization of a particular object, such as an engine, (2) the optimization of a particular method, such as a construction process, (3) the optimization or detection of a state of lowest energy, such as a physical problem, or (4) the optimization of a solution to any search problem, such as the problem of satisfiability or the traveling salesman problem.

More formally, a simple binary optimization problem is defined for a problem space S that is characterized by a bit string of a certain length l: $S = \{0, 1\}^l$. Each bit string represents a particular problem solution. Feedback is provided in terms of a scalar reward (fitness) value that rates the solution quality. An optimization algorithm should be designed to *effectively* search the solution space for the global optimum.

Given, for example, the problem of optimizing a smoothie drink with a choice of additional ingredients mango, banana, and honey available, the problem may be coded by three bits indicating the absence (0) or presence (1) of each ingredient. The fitness is certainly very subjective in this example, but assuming that we like all three ingredients equally well, prefer any combination of two of the ingredients over just one ingredient alone, and like the combination of all three ingredients best, we constructed a *one-max problem*.

Table 2.1 (first numeric column) gives possible numerical values for a four bit one-max problem. The best solution to the problem is denoted by 1111 and the fitness is determined by the number of ones in the problem. Certainly, the one-max problem is a very easy problem. There is only one (global) optimum and the closer to the global optimum, the higher the fitness. Thus, the problem provides strong *fitness guidance* towards the global optimum. This guidance can be even more clearly observed in the six bit visualization of the one-max problem shown in Figure 2.1a: The more ones that are specified in a solution, the higher the fitness of the solution.

However, problems can be *misleading* in that the fitness measure may give incorrect clues in which direction to search for more promising solutions. Table 2.1 and Figure 2.1 show progressively more misleading types of problems. While the *royal-road* function still provides the steps that lead to the best solution (Mitchell, Forrest, & Holland, 1991), the *needle in the haystack* problem provides no directional clues whatsoever—only the optimum results in high fitness. The *trap problem* provides quality clues in the opposite direction: Fitness leads towards a local optimum away from the global optimum. In a trap problem the combination of a number of factors (e.g. ingredients) makes the solution better, but the usage of only parts of these factors makes the solution actually worse than the base solution that uses none.

The reader should not be misled by these abstract problems thinking that a local optimum (if existent) and the global optimum will always be the exact inverse of a binary string. In other, albeit related problems, this certainly does not need to be the case. Finally, the meaning of zeroes and ones may be (partially) swapped so that the global optimum does not need to be all ones but the problem structure may still be identical to one of the types outlined in the table. Thus, although the examples are very simplified, they characterize important problem structures that pose different challenges to the learning algorithm.

The problem substructures may be combined to form larger, more complex problems. Each substructure can then be regarded as a building block (BB) in the larger problem, that is, a substructure that is evaluated rather independently of the other structures. Certainly, there can be higher-order dependencies or overlapping BB structures. Also, the fitness contribution of each BB may vary.

Table 2.1. There are several ways in which the solution quality measure can lead towards the global optimal solution (here: 1111). In the *one-max* problem, the closer the solution is to the optimal solution, the higher its quality. In the *royal-road* problem, the path leads to the optimal solution step-wise. In the *needle in the haystack* problem, fitness gives no hints about the optimal solution. In the *trap problem*, the quality measure is *misleading* since the more the solution differs from the optimal solution, the higher its fitness.

	One-Max	Royal-Road	Needle i.H.	Trap Prob.
0000	0	0	0	3
0001	1	0	0	2
0010	1	0	0	2
0100	1	0	0	2
1000	1	0	0	2
0101	2	0	0	1
1001	2	0	0	1
0110	2	0	0	1
1010	2	0	0	1
0011	2	2	0	1
1100	2	2	0	1
0111	3	2	0	0
1011	3	2	0	0
1101	3	2	0	0
1110	3	2	0	0
1111	4	4	4	4

Two most basic combinations can be distinguished. In the simplest case the fitness contribution of each substructure is equal and simply added together yielding an *uniformly scaled* problem. Another method is to exponentially scale the utility of each substructure. In this case the fitness of the second block matters only once the optimum of the first block is found and so forth, yielding an *exponentially scaled* problem. It should be noted that in a uniformly scaled problem the blocks can be solved in parallel because fitness provides information about all blocks. On the other hand, in an exponentially scaled problem, the BBs need to be solved sequentially since the exponentially scaled fitness nearly eliminates fitness information from later blocks that are not most relevant yet.

Problem structures that are composed of many smaller dependency structures are often referred to as *decomposable problems* or problems of bounded difficulty (Goldberg, 2002). Boundedly difficult problems are bounded in that the BB size of lower-level interactions is bounded to a certain *order*—the order of problem difficulty. For example, from the problems in Table 2.1, the one-max problem has an order of difficulty of one because it is actually composed of four uniformly scaled BBs of size one. The royal-road function is of order two since two BBs are combined. Both the needle in the haystack and

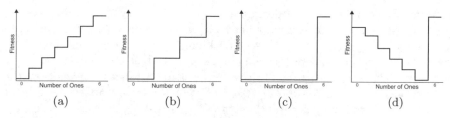

Fig. 2.1. The figures show progressively more challenging optimization problems. In the one-max problem (a), fitness progressively increases on the way to the globally optimal solution (here: all ones). In the royal-road problem (b), larger steps need to be taken towards the optimal solution. In the needle in the haystack problem (c), fitness gives no information about where the optimal solution is situated. Finally, in the trap problem (d), fitness actually misleads to a local optimum (here: all zeroes) and thus away from the globally optimal solution (here: all ones).

the trap problem are of order four since there is no further decomposition possible.

Regardless of the problem structure, the problem may actually have multiple optimal solutions, the quality measure may be noisy, or the provision of several near-optimal solutions may be more desirable than the detection of one (completely) optimal solution. Often, an expert may want to choose from such a set of (near-) optimal solutions. In this case, a learner would be required to find not only one globally optimal solution but rather a set of several different (near-) optimal solutions.

To summarize, optimization problems are problems in which a best solution must be found given a solution space. Feedback is available that rates solutions proposed by the learner. The feedback may or may not provide hints where to direct the further search for the optimal solution. Finally, the number of optimal solutions may vary and, dependent on the problem, one or many optimal (or near optimal) solutions may need to be found.

2.1.2 Classification Problems

A classification problem poses further difficulties to the learning algorithm. Although a classification problem may be reduced to an optimization problem, the reduction is tedious and often destroys much of the available problem structure and information inherent in a classification problem.

We define a classification problem as a problem that consists of problem instances $s \in \mathcal{S}$. Each problem instance belongs to one class (traditionally termed an action in LCSs) $a \in \mathcal{A}$. In machine learning terms, s may be termed a feature vector and a a concept class. The mapping from $\mathcal{S}$ to $\mathcal{A}$ is represented by a *target concept* belonging to a set of concepts (that is, the *concept space*). The goal of a classification system is to learn the target concept. Thus, the classification system learns to which class a_i each problem instance s_i belongs to. The desirable properties of such a learning system are that the learner learns a maximally *accurate* problem solution, measured usually by the percentage of correct problem instance classifications, and a maximally *general* problem solution, which can be characterized as a solution that generalizes well to other (unseen) problem instances. Given that the learner has a certain *hypothesis space* of expressible solutions, the learner looks for the maximally accurate, maximally general hypothesis with respect to the target concept.

As in optimization problems, a classification problem may be composed of several subproblems characterizable as BBs. However, such BBs cannot be directly related to fitness but can only increase the probability that a problem instance belongs to a certain class. Solution hypotheses may be represented in a more distributed fashion in that different subsolutions may be responsible for different problem subspaces. This is expected to be particularly useful if different problem subspaces are expected to have a quite different class distribution. In this case, different BBs will be relevant dependent on the current problem subspace. Thus, in contrast to optimization problems, different BBs may need to be detected and propagated in different problem subspaces. Nonetheless, the global BB distribution can also be expected to yield important information that can improve the search for accurate problem (sub-)solutions.

Boolean Function Problems

In most of this work, we focus on Boolean function problems. In these problems, the problem instance space is restricted to the binary space, that is, $\mathcal{S} \subseteq \{0,1\}^l$ where l denotes the fixed problem length. Similarly, a Boolean function problem has only two output classes $\mathcal{A} = \{0,1\}$. Consequently, any Boolean function can be represented by a logical formula and consequently also by a logical formula in disjunctive normal form (DNF). Appendix C introduces the Boolean function problems investigated herein showing an exemplar DNF representation and discussing their general structure and problem difficulty.

As an example, let us consider the well-known multiplexer problem, which is widely studied in LCS research (De Jong & Spears, 1991; Wilson, 1995; Wilson, 1998; Butz, Kovacs, Lanzi, & Wilson, 2001). It has been shown that LCSs are superior compared to standard machine learning algorithms, such as the decision tree learner $C4.5$, in the multiplexer task (De Jong & Spears, 1991). The problem is of particular interest due to its dependency structure and its distributed niches, or subsolutions. The problem is defined for binary strings of length $l = k + 2^k$. The output of the multiplexer function is determined by the bit situated at the position referred to by k position bits (usually but not necessarily located at the first k positions). The disjunctive normal form of the 6-multiplexer for example can be written as follows:

$$6MP(x_1, x_2, x_3, x_4, x_5, x_6) = \neg x_1 \neg x_2 x_3 \vee \neg x_1 x_2 x_4 \vee x_1 \neg x_2 x_5 \vee x_1 x_2 x_6; \quad (2.1)$$

for example, $f(100010) = 1$, $f(000111) = 0$, or $f(110101) = 1$. More information on the multiplexer problem can be found in Appendix C. It is interesting to see that the DNF form of the multiplexer problem consists of conjunctions that are *non-overlapping*. Any problem instance belongs, if at all, to only one of the conjunctive terms in the problem. Later, we will see that the amount of overlap in a problem is an important problem property for LCSs' learning success.

Real-valued Problems

Boolean function problems are a rather restricted class of classification problems. In the general case, a problem instance s may be represented by a feature vector. Each feature may be a binary attribute, a nominal attribute, an integer attribute, or a real-valued attribute. Mixed representations are possible.

We refer to such real-world classification problems as datamining problems. The problem is represented by a set of problem instances with their corresponding class. A problem instance may consist of a mixture of features and there may be more than two classification classes. Since the target concept is generally unknown in datamining problems, performance of the learner is often evaluated by the means of stratified ten-fold cross-validation (Mitchell, 1997) that trains the system on a subset of the data set and tests it on the remaining problem instances. The data is partitioned into ten subsets. The learner is trained on nine of the ten subsets and tested on the remaining subset. To avoid sampling biases, this procedure is repeated ten times, each time training and testing on different subsets. Stratification assures that the class distribution is approximately equal in all folds. Ten-fold cross-validation is very useful in evaluating the generalization capabilities of the learner since performance is tested on previously unseen data instances.

2.1.3 Reinforcement Learning Problems

In contrast to optimization and classification problems, feedback in the form of a class or immediate reinforcement is not necessarily available immediately in RL problems. Rather, feedback is provided in terms of a scalar reinforcement value that indicates the quality of a chosen action (or classification). Additionally, successive problem instances may be dependent upon each other in that subsequent input usually depends on previous input and on the executed actions. RL problems are thus more difficult, but also more natural, simulating interaction with an actual outside world. Figure 2.2 shows the agent-environment interaction typical in RL problems.

Despite the environmental interaction metaphor, a classification problem may be redefined as an RL problem providing reward feedback about the accuracy of the chosen class. For example, a reward of 1000 may be provided for the correct class and a reward of 0 for the incorrect class. In this case reward is not delayed. We refer to such redefined classification problems as

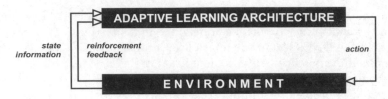

Fig. 2.2. In RL problems an adaptive agent interacts with an environment executing actions and receiving state information and reinforcement feedback.

single-step RL problems. On the other hand, *multistep RL problems* refer to RL problems in which reward is delayed and successive states depend on each other and on the chosen action. In the latter case, reward (back-)propagation is necessary.

Later, we see that LCSs are online generalizing RL mechanisms. Classification problems are usually converted into single-step RL problems when learned by LCSs. For convenience reasons we usually refer to these single-step RL problems as classification problems. However, the reader should keep in mind that when referring to classification problems in conjunction with an LCS application, the LCS actually faces a single-step RL problem.

Two types of multistep RL problems need to be distinguished: those that are modeled by a Markov decision process (MDP) and those that are modeled by a partially observable Markov decision process (POMDP).

Markov Decision Processes

We define a multistep problem as a Markov decision process (MDP), reusing notation from the classification problems where appropriate. An MDP problem consists of a set of possible sensory inputs $s \in \mathcal{S}$ (i.e. the states in the MDP); a set of possible actions $a \in \mathcal{A}$; a state transition function $f : \mathcal{S} \times \mathcal{A} \rightarrow \Pi(\mathcal{S})$, where $\Pi(\mathcal{S})$ denotes a probability distribution over all possible next states ($\mathcal{S}$); and a reinforcement function $\mathcal{R} : \mathcal{S} \times \mathcal{A} \times \mathcal{S} \rightarrow \Re$. The state transition function defines probabilities for reaching all possible next states given the current state and the current action. The reinforcement function defines the resulting reward, which depends on the current state transition. For example, at a certain point in time t, state s_t may be given. The system may then decide on the execution of action a_t. The execution of a_t leads to the reception of reward r_t and the perception of the consequent state s_{t+1}.

An MDP is called a *Markov* decision process because it satisfies the Markov property: Future probabilities and thus the optimal action decision can be determined from the current state information alone since the state transition probabilities depend solely on the current state information.

A simple example of a (multistep) MDP problem is Maze 1—a simple maze environment shown in Figure 2.3. The learning system, or *agent*, may

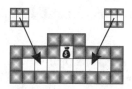

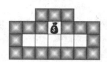

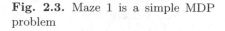

Fig. 2.3. Maze 1 is a simple MDP problem

Fig. 2.4. The two identical looking states in Maze 2 turn the problem into a POMDP problem.

reside in one of the five positions in the maze perceiving the eight surrounding positions. An empty position may be coded by 0, a blocked position by 1. The money bag indicates a reward position in which reward is received and the agent is reset to a randomly chosen empty position. Note that each position has a unique perception code so that the problem can be modeled by an MDP process.

Partially Observable Markov Decision Processes

More difficult than MDP problems, are POMDP problems in which current sensory input may not be sufficient to determine the optimal action. Formally, a POMDP can be defined by a state space $\mathcal{X}$, a set of possible sensations $\mathcal{S}$, a set of possible actions $\mathcal{A}$, a state transition function $f : \mathcal{X} \times \mathcal{A} \rightarrow \Pi(\mathcal{X})$. In contrast to an MDP problem, however, states are not perceived directly but are converted into a sensation using an observation function $O : \mathcal{X} \rightarrow \Pi(\mathcal{S})$ that converts a particular state into a sensation. Similarly, the reward function does not rely on the sensations but on the underlying (unobservable) states $R : \mathcal{X} \times \mathcal{A} \times \mathcal{X} \rightarrow \Re$. In contrast to the MDP, a POMDP might violate the Markov property in that optimal action decisions cannot be made solely based on current sensory input.

A simple example of a POMDP problem is shown in Figure 2.4. Although only slightly larger than Maze 1, this maze does not satisfy the Markov property since the second empty position on the left looks identical to the second empty position on the right. Thus, given that the agent is currently in either of the two positions, it is impossible to know if the fastest way to the reward is to the left or to the right. Only an internal state or a short-term memory (that may, for example, indicate that the agent came from the left-most empty position) can disambiguate the two states and allow the agent to act optimally in the positions in question.

In the studies herein, we focus on MDP problems. Applications of the learning classifier system XCS to POMDP problems can be found elsewhere (Lanzi & Wilson, 2000; Lanzi, 2000), in which the system is enhanced with internal memory mechanisms somewhat similar to Holland's original message list idea.

2.2 Reinforcement Learning

Facing an MDP or POMDP problem, an RL system is the most appropriate system to solve the problem. In essence, the investigated rule-based evolutionary systems are RL systems that use GAs to evolve their state-action-value representation. Excellent introductions to RL are available (see Kaelbling, Littman, & Moore, 1996; Dietterich, 1997; Sutton & Barto, 1998) and the following overview can only provide a brief glance at RL. We introduce terminology and basic understanding necessary for the remainder of the book.

The task of an RL system is to learn an optimal behavioral policy interacting with an MDP (or POMDP) problem. A behavioral policy is a policy that decides on action execution. Given current sensory input (and possibly further internal state information) the behavioral policy is responsible for the action decision. A behavioral policy is optimal if it results in the maximum expectable reward in the long run. The most often used expression to formalize the expected reward is the *cumulative discounted reward*:

$$E(\sum_{t=0}^{\infty} \gamma^t r_t),$$ (2.2)

where $\gamma \in [0, 1]$ denotes the discount factor that weighs the importance of more distant rewards. Setting γ to zero results in a single-step RL problem in which only current reward is important. Setting γ to one results in a system in which cumulative reward needs to be optimized. Usually, γ is set to values close to one such as 0.9. RL essentially searches for a behavioral policy that maximizes the cumulative discounted reward.

Looking back at our small maze problem in Figure 2.3, we can see how much reward can be expected in each state executing an optimal behavioral policy. Assuming that the environment triggers a reward of 1000 when the rewarding position is reached, the three positions that are one step away from the reward position have an expected reward of 1000 and the two outermost positions have an expected reward of 900.

RL systems learn state value or state-action value representations using temporal difference learning techniques to estimate and maximize the expected discounted reward expressed in Equation 2.2. Two approaches can be distinguished: (1) *Model-free* learners learn an optimal behavioral policy directly without learning the state-transition function; (2) *Model-based* learners learn the state-transition function using it to learn or improve their behavioral policy.

2.2.1 Model-Free Reinforcement Learning

Two major approaches comprise the model-free RL realm: (1) $TD(\lambda)$ and (2) *Q-learning*. While the former needs an interactive mechanism that updates the RL-based critic and the behavioral policy in turn, the latter is doing the same more naturally. After a short overview of $TD(\lambda)$ we focus on Q-learning since it is able to learn an optimal policy *off-policy* (that is, independent of the current behavioral policy). The RL mechanism implemented in the subsequently investigated learning classifier system XCS closely resembles Q-learning.

$TD(\lambda)$ methods interactively, or in turn, update their current behavioral policy π and the critic V^π that evaluates the policy π. Given the current critic V^π, a k-armed bandit optimization mechanism may be used to optimize π. Given current policy π, V^π may be updated using the $TD(\lambda)$ strategy. Hereby, each state value $V(s)$ is updated using

$$V(s) \leftarrow V(s) + \beta(r + \gamma V(s') - V(s))e(s), \qquad (2.3)$$

$$e(s) = \sum_{k=1}^{t} (\lambda\gamma)^{t-k} \delta_{s,s_k}, \text{ where } \delta_{s,s_k} = \begin{cases} 1 & \text{if } s = s_k \\ 0 & \text{otherwise} \end{cases}$$

where $e(s)$ denotes the eligibility of state s, meaning its involvement in the achievement of the current reward. Parameter β denotes the learning rate somewhat reflecting the belief in the new information vs. the old information. A large β assumes very low prior knowledge resulting in a large change in the value estimate, whereas a small β assumes solid knowledge resulting in a small change in the value estimate. Parameter λ controls the importance of states in the past and δ monitors the occurrences of the states. With $\lambda = 0$, past states are not considered and only the currently encountered state transition is updated. Similarly, setting λ to one, all past states are considered as equally relevant. Note that the update is still discounted by γ so that the influence of parameter λ specifies the belief in the relevancy of the current update for the states encountered in the past (Kaelbling, Littman, & Moore, 1996).

A somewhat more natural approach is Watkins' Q-learning mechanism (Watkins, 1989). Instead of learning state values, Q-learning learns state-action values, effectively combining policy and critic in one. A *Q-value* of a state action pair (s, a) essentially estimates the expected discounted future reward if executing action a in state s and pursuing the optimal policy thereafter. Q-values are updated by

$$Q(s,a) \leftarrow Q(s,a) + \beta(r + \gamma \text{ max}_{a'} Q(s',a') - Q(s,a)). \qquad (2.4)$$

Due to the `max` operator, the Q-value predicts the average discounted reward of executing a in state s reaching state s' and following the optimal policy thereafter. Q-learning is guaranteed to converge to the optimal values as long as it is guaranteed that in the long run all state-action pairs are executed infinitely often and learning rate β is decayed appropriately. If the optimal Q-values are determined, the optimal policy is determined by

$$\pi^*(s) = \text{arg max}_a Q(s,a), \qquad (2.5)$$

which chooses the action that is expected to maximize the consequent Q-value if executed in current state s. To ensure an infinite exploration of the state-action space, an ϵ-greedy exploration strategy may be used:

$$\pi(s) = \begin{cases} \texttt{arg max}_a Q(s,a) & \text{with probability } 1-\epsilon \\ \texttt{rand}(a) & \text{otherwise} \end{cases}, \qquad (2.6)$$

which chooses a random action with probability ϵ.

In our running Maze 1 example, the Q-table learned by a Q-learner is shown in Table 2.2. The environment provides a constant reward of 1000 when the money position is reached. Additionally, after reward reception, the agent is reset to a random position in the maze. The reward is propagated backward through the maze yielding lower reward to more distant positions. In effect, Q-learning is an online distance learning mechanism to a reward source where the reward prediction, or *Q-value*, indicates the worthiness or value of executing an action given the current situation. Note that worthiness in Q-learning is defined as the expected discounted future reward using Equation 2.4. Other discount mechanisms often work just as well depending on the task setup. For example, parameter γ could be set to one but actions may have an associated, potentially fixed, cost value which will be deducted from the expected future reward. In this case, explicit discounting is unnecessary since the environment—actually reflecting the inner architecture of the agent—would take care of discounting. The cost would reflect a potentially body-specific effort measure of the executed action.

Table 2.2. The Q-values in the shown Q-table reflect the value of each available action in each possible state of Maze 1 (shown on the left). Each value is defined as the immediate reward plus the expected discounted future reward received after executing the action indicated by the arrows in the state defined by the letters. The corresponding sensation is derived by specifying the surrounding conditions starting north and coding clockwise, specifying an obstacle by symbol 1 and a free space by symbol 0. The discount level is set to $\gamma = .9$.

	stat e sensation	↑	↗	→	↘	↓	↙	←	↖
A	11011111	810	810	900	810	810	810	810	810
B	10011101	900	1000	900	900	900	900	810	900
C	01011101	1000	900	900	900	900	900	900	900
D	11011100	900	900	810	900	900	900	900	1000
E	11111101	810	810	810	810	810	810	900	810

To summarize, Q-learning learns a Q-function that determines state-dependent value estimates for each available action in each state of the environment. Due to its well-designed interactive learning mechanism and proven convergence properties, it is commonly used in RL. As we will see, Q-learning forms the fundamental basis for the RL component used in the investigated LCSs.

2.2.2 Model-Based Reinforcement Learning

In addition to the reward prediction values, model-based RL techniques learn (potentially an approximation of) the state-transition function of the underlying MDP problem. Once the state-transition function is learned with sufficient accuracy, an optimal behavioral policy can be learned by simply simulating state-transitions offline using *dynamic programming* techniques (Bellman, 1957).

In dynamic programming, the state transition function is known to the system and used to estimate the payoff for each state-action pair. The literature on dynamic programming is broad, conveying many convergence results concerning different environmental properties such as circles, probabilistic state transitions, and probabilistic reward (see e.g. Bellman, 1957; Gelb, Kasper, Nash, Price, & Sutherland, 1974; Sutton & Barto, 1998).

In model-based RL, the state-transition function needs to be learned as suggested in Sutton's DYNA architecture (Sutton, 1990). DYNA uses state-transition experiences in two ways: (1) to learn a behavioral policy (using $TD(\lambda)$ or Q-learning); (2) to learn a predictive model of the state-transition function. Additionally, DYNA executes offline policy updates (independent of the environmental interactions) using the learned predictive model. Moore and Atkeson (1993) showed that the offline learning mechanism can be sped up significantly if the offline learning steps are executed in a prioritized fashion favoring updates that promise to result in large state(-action) value changes. DYNA and related techniques usually represent their knowledge in tabular form.

Later, we will see that GAs can be applied in LCSs to learn a more generalized representation of Q-values and predictions. LCSs essentially try to cluster the state space so that each cluster accurately predicts a certain value. GAs are used to learn the appropriate clusters.

2.3 Genetic Algorithms

Prior to LCSs, John H. Holland proposed GAs (Holland, 1971; Holland, 1975). Somewhat concurrently, evolution strategies (Rechenberg, 1973; Bäck & Schwefel, 1995) were proposed, which are very similar to GAs but are not discussed any further herein. Goldberg (1989) provides a comprehensive introduction to GAs including LCSs. Goldberg's recent book (Goldberg, 2002) provides a much more detailed analysis of GAs including scale-up behavior as well as problem and parameter/operator dependencies.

This section gives a short overview of the most important results and basic features of GAs. The interested reader is referred to Goldberg (1989) for a comprehensive introduction and to Goldberg (2002) for a detailed analysis of GAs leading to the design of *competent GAs*—GAs that solve boundedly difficult problems quickly, accurately, and reliably.

2.3.1 Basic Genetic Algorithm

GAs are evolutionary-based search or optimization techniques. They evolve a *population* (or set) of individuals, which are represented by a specific *genotype*. The decoded individual, also called *phenotype*, specifies the meaning of the individual. For example, when facing the problem of optimizing the ingredients of a certain dish, an individual may code the presence (1) or absence (0) of the available ingredients. The decoded individual, or phenotype, would then be the actual ingredients put together.

Basic GAs face an optimization problem as specified in Section 2.1.1. Feedback is provided in form of a fitness value that specifies the quality of an individual, usually in the form of a scalar value. If we face a maximization problem, a high fitness value denotes high quality. GAs are designed to progressively generate (that is, evolve) individuals that yield higher fitness.

Given a population of evaluated individuals, *evolutionary pressures* are applied to the population generating offspring and deleting old individuals. A basic GA comprises the following steps executed in each iteration given a population:

1. evaluation of current population
2. selection of high fitness individuals from current population
3. mutation and recombination of selected individuals
4. generation of new population

The evaluation is necessary to derive a fitness measure for each individual. Selection focuses the current population on more fit individuals, which represent more accurate subsolutions. Mutation is a diversification operator that searches in the syntactic, genotypic neighborhood of an individual by slightly changing its structure. Recombination, also called *crossover*, recombines the structure of parental individuals in the hope of generating offspring that combines the positive structural properties of both parents. Finally, the generation of the new population decides which individuals are part of the new population. In the simplest case, the number of reproduced classifiers equals the population size and the offspring individuals replace the old individuals. In a more steady-state version, less individuals are reproduced and only a subset of the old population is replaced. Replacement may also depend on fitness.

The remainder of this chapter analyzes the different methods in more detail. Hereby, we follow Goldberg's facetwise GA theory.

2.3.2 Facetwise GA Theory

To understand a system as complex as a GA, it is helpful to partition the system into its most relevant components and investigate those components separately Once the single components are sufficiently well understood, they may then be combined appropriately, respecting interaction constraints. This is the essential idea behind Goldberg's facetwise approach to GA theory (Goldberg, 1991; Goldberg, Deb, & Clark, 1992; Goldberg, 2002).

Before we can do the partitioning, though, it is necessary to understand *how* a GA is supposed to work and *what* it is supposed to learn.

Since GAs are targeted to solve optimization problems evolving the solution that yields highest fitness, fitness is the crucial factor along the way to an optimal solution. Learning success in GAs relies on sufficient *fitness guidance* towards the optimal solution. However, fitness may be misleading as illustrated above in the trap problem (Table 2.1). Thus, the structural assumption made in GAs is that of *bounded difficulty*, which means that the overall optimization problem is composed of substructures, or BBs, that are of bounded length. The internal structure of one BB might be misleading in that a BB might, for example, resemble a trap problem. However, the overall problem is assumed to provide some fitness guidance so that good combinations of BBs are expected to lead towards the optimal solution.

With this objective in mind, it is clear that GAs should detect and propagate subproblems effectively. The goal is to design *competent GAs* (Goldberg, 2002)—GAs that solve boundedly difficult problems quickly, accurately, and reliably. Hereby, quickly means to solve the problems in low-order polynomial time (ideally subquadratic), with respect to the problem length. Accurately means to find a solution in small fitness distance from the optimal solution, and reliably means to find this solution with a low error probability.

The actual GA design theory (Goldberg, 1991; Goldberg, Deb, & Clark, 1992; Goldberg, 2002) then stresses the following points for GA success:

1. Know what GAs process: Building blocks
2. Know the GA challenge: BB-wise difficult problems
3. Ensure an adequate supply of raw BBs
4. Ensure increased market share for superior BBs
5. Know BB takeover and convergence times
6. Make decisions well among competing BBs
7. Mix BBs well

While we characterized building blocks and BB-wise difficult problems (that is, boundedly difficult problems) above, we elaborate on the latter points in the subsequent paragraphs. First, we focus on supply and increased market share. Next, we look at diversification, BB decision making, and effective BB mixing.

2.3.3 Selection and Replacement Techniques

As in real-live Darwinian evolution (Darwin, 1859), selection, reproduction, and deletion decide the life and death of individuals and groups of individuals in a population. Guided by fitness, individuals are evolved that are more fit for the problem at hand. Several selection and deletion techniques exist, each with certain advantages and disadvantages. For our purposes most important are (1) *proportionate selection* (also often referred to as *roulette-wheel selection*) and (2) *tournament selection*. Most other commonly used selection mechanisms are comparable to either of the two. More detailed comparisons of different selection schemes can be found elsewhere (Goldberg, 1989; Goldberg & Deb, 1991; Goldberg & Sastry, 2001).

Proportionate Selection

Proportionate selection is the most basic and most natural selection mechanism: Individuals are selected for reproduction proportional to their fitness. The higher the fitness of an individual, the higher its probability of being selected. In effect, high fitness individuals are evolved. Given a certain individual with fitness f_i and an overall average fitness in the population $\overline{f}$, the proportion of individual p_i is expected to change as

$$p_i \leftarrow \frac{f_i}{\overline{f}} p_i \tag{2.7}$$

A similar equation can be derived for the proportion of a BB representation and its fitness contribution in the population (Goldberg & Sastry, 2001). The equation shows that structural growth by the means of proportionate selection strongly depends on *fitness scaling* and the current *fitness distribution* in the population (Baker, 1985; Goldberg & Deb, 1991; Goldberg & Sastry, 2001).

A simple example clarifies the dependence on fitness scaling. Given any fitness measure that needs to be maximized, the probability of reproducing the best individual is the following:

$$P(\texttt{rep. of best individual}) = \frac{f_{max}}{\overline{f}} p_i \tag{2.8}$$

Assuming now that we add a positive value x to the fitness function, the resulting probability of reproduction decreases to

$$P(\texttt{rep. of best individual scaled}) = \frac{f_{max} + x}{\overline{f} + x} p_i \tag{2.9}$$

The larger x, the smaller the probability that the best individual is selected for reproduction until all selection probabilities are equal to their current proportion, although the best individual should usually receive significantly more reproductive opportunities. The manipulation is certainly inappropriate but shows that a fitness function needs to be well-designed—or appropriately scaled—to ensure sufficient selection pressure when using proportionate selection.

The dependency on the current fitness distribution has a strong impact on the convergence to the global best individual of a population. The more similar the fitness values in a population, the less selection pressure is encountered by the individuals. Thus, once the population has nearly converged to the global optimum, fitness values tend to be similar so that selection pressure due to proportionate selection is weak. This effect is undesirable if a single global solution is searched.

However, complete convergence may not necessarily be desired or may even be undesired if multiple solutions are being searched for. In this case, proportionate selection mechanisms might actually be appropriate given a reasonable fitness value estimate. For example, as investigated elsewhere (Horn, 1993; Horn, Goldberg, & Deb, 1994), proportionate selection can guarantee that all similarly good solutions (or subsolution niches) can be sustained with a population size that grows linearly in the number of niches and logarithmically in the time they are assured to be sustained. The only requirement is that fitness sharing techniques are applied and that the different niches are sufficiently non-overlapping.

Tournament Selection

In contrast to proportionate selection, tournament selection does not depend on fitness scaling (Goldberg & Deb, 1991; Goldberg, 2002). In tournament selection, tournaments are held among randomly selected individuals. The tournament size s specifies how many individuals take part in a tournament. The individual with the highest fitness wins the tournament and consequently undergoes mutation and crossover and then serves as a candidate for the next generation.

If replacing the whole population by individuals selected by tournament selection, the best individuals can be expected to be selected s times so that the proportion p_i of the best individual i can be expected to grow with s, that is:

$$p_i \leftarrow sp_i \qquad (2.10)$$

That means that the best individual is expected to take over the population quickly since the proportion of the best individual grows exponentially in s. Thus, in contrast to proportionate selection, which naturally stalls late in the run, tournament selection pushes the better individuals until the best available individual takes over the whole population.

Supply

Before selection can actually be successful, BBs need to be available in the population. This leads to the important issue of initial BB *supply*. If the initial population is too small to guarantee the presence of all important BBs (expressed in usually different) individuals, a GA relies on mutation to generate the missing structures by chance. However, an accidental successful mutation is very unlikely (exponentially decreasing in the number of missing BB values). Thus, a sufficiently large population with an initially sufficiently large diversity is mandatory for GA success.

Niching

Very important for a successful application of GAs in the LCS realm is the parallel sustenance of equally important subsolutions. Usually, *niching* techniques are applied to accomplish this sustenance. Hereby, two techniques reached significant impact in the literature: (1) crowding, and (2) sharing.

In crowding (De Jong, 1975; Mahfoud, 1992; Harik, 1994) the replacement of classifiers is restricted to classifiers that are (usually syntactically) similar. For example, in the restricted tournament selection technique (Harik, 1994), offspring is compared with a subset of other individuals in the population. The offspring competes with the (syntactically) closest individual, replacing it, if its fitness is larger.

In sharing techniques (Goldberg & Richardson, 1987) fitness is shared among similar individuals where similarity is defined by an appropriate distance measure (e.g. Hamming distance in the simplest case). The impact of sharing has been investigated in detail (Horn, 1993; Horn, Goldberg, & Deb, 1994; Mahfoud, 1995). Horn, Goldberg, and Deb (1994) highlight the importance of fitness sharing in the realm of LCSs showing the important impact of sharing on the distribution of the population and potentially near infinite niche sustenance due to the applied sharing technique.

Although sharing can be beneficial in non-overlapping (sub-)solution representations, the more the solutions overlap, the less beneficial fitness sharing techniques become. Horn, Goldberg, and Deb (1994) propose that sharing works successfully as long as the overlap proportion is smaller than the fitness ratio between the competing individuals. Thus, if the individuals have identical fitness, only a complete overlap eliminates the sharing effect. In general, the higher the degree of overlap, the smaller the sharing effect and thus the higher the probability of losing important BB structures due to genetic drift.

2.3.4 Adding Mutation and Recombination

Selection alone certainly does not do much good. In essence, selection uses a complicated method to find the best individuals in a given set of individuals. Certainly, this can be done much faster by simple search mechanisms. Thus, selection needs to be combined with other search techniques that search in the *neighborhoods* of the current best individuals for even better individuals. The two basic operators that accomplish such a search in GAs are *mutation* and *crossover*.

Simple mutation takes an individual as input and outputs a slight variation of the individual. In the simplest form when coding an individual in binary, mutation randomly flips bits in the individual with a certain probability μ. Effectively, mutation searches in the *syntactic* neighborhood of the individual where the shape and size of the neighborhood depend on the mutation operator and the mutation probability μ.

Crossover is designed to recombine current best individuals. Thus, rather than searching in the syntactic neighborhood of one individual, crossover searches in the neighborhood defined by two individuals, resulting in a certain type of knowledge exchange among the crossed individuals. In the simplest case when coding individuals in binary, *uniform crossover* exchanges each bit with a 50% probability, whereas one-point or two-point crossover choose one or two positions in the bit strings and exchange the right or the inner part of the resulting partition, respectively.

It should be noted that uniform crossover does not assume any relationship among bit positions, whereas one- and two-point crossover implicitly assume that bits that are close to each other depend on each other since substrings are exchanged. One-point crossover additionally assumes that beginning and ending are unrelated to each other, whereas two-point crossover assumes a more circular coding structure. More detailed analyses of simple crossover operators can be found elsewhere (Bridges & Goldberg, 1987; Booker, 1993).

It is important to appreciate the effects of mutation and crossover alone, disregarding selection for a moment. If selecting randomly and simply mutating individuals, mutation causes a general *diversification* in the population. In the long run, each attribute in an individual will be set independently uniformly distributed resulting in a population with maximum entropy in its individuals. In combination with selection, mutation causes a search in the syntactic neighborhood of an individual where the spread of the neighborhood is controlled by the mutation type and rate. Mutation may be biased, incorporating potentially available problem knowledge to improve the neighborhood search as well as to obey problem constraints.

Recombination, on the other hand, exchanges information among individuals syntactically dependent on the structural bias in the applied crossover operator. Selecting randomly, random crossover results in randomized shuffling of the individual genotypes. In the long run, crossover results in a random distribution of attribute values over the classifiers but does not affect the proportion of each value in the population.

In conjunction with selection, crossover is responsible for effective BB processing. Crossover needs to be designed to ensure proper BB recombination by exchanging important BBs among individuals and preventing the disruption of BBs in the meantime. Goldberg (2002) compares this very important exchange of individual substructures with innovation. Since innovation essentially refers to a successful (re-)combination of available knowledge in a novel manner, GAs are essentially designed (or should be designed) to do just that—be innovative in a certain sense.

Unfortunately, standard crossover operators are not guaranteed to propagate BBs effectively. BBs may also be disrupted by destroying important BB structures when recombining individuals. Dependent on the crossover operator and BB structure, a probability may be derived that the BB is not fully exchanged but cut by crossover. If it is cut and only a part is exchanged, the

BB structure may get lost dependent on the structure present in the other individual.

To prevent such disruption and design a more directed form of innovation, estimation of distribution algorithms (EDA) were recently introduced to GAs (Pelikan, Goldberg, & Lobo, 2002; Larrañaga, 2002). These algorithms estimate the current structural distribution of the best individuals in the population and use this distribution estimation to constrain the crossover operator or to generate better offspring directly from the distribution. In Chapter 7, we show how to incorporate mechanisms from the extended compact GA (ECGA) (Harik, 1999), which learns a non-overlapping BB structure, as well as from the Bayesian optimization algorithm (BOA), which learns a Bayes model of the BB structure, into the investigated XCS classifier system.

2.3.5 Component Interaction

We already discussed that a combination of selection with mutation and recombination leads to local search plus potentially innovative search. In addition to the impact of selection on growth and convergence, selection strength is interdependent with mutation and recombination. The two methods interact in that selection propagates better individuals while mutation and crossover search in the neighborhood of these individuals for even better solutions. Consequently, the growth of better individuals (or better BBs), often characterized by a *take over time* (Goldberg, 2002), needs to be balanced with the search in the neighborhood of the current best solutions. Too strong selection pressure may result in a collapse of the population to one only locally optimal individual, preventing effective search and innovation via mutation and crossover. On the other hand, too weak selection pressure may allow *genetic drift* that can cause the loss of important BB structures by chance.

These insights led to the proposition of a *control map* for GAs (Goldberg, Deb, & Thierens, 1993; Thierens & Goldberg, 1993; Goldberg, 1999) that characterizes a region of selection and recombination parameter settings in which a GA can be expected to work. The region is bounded by *drift*, when selection pressure is too low, *cross-competition*, when selection is too high, and *mixing*, when knowledge exchange is too slow with respect to the selection pressure. The mixing bound essentially characterizes the boundary below which knowledge exchange caused by crossover is not strong enough with respect to the selection pressure applied. Essentially the time until the expected generation of a better individual needs to be shorter than the mentioned take over time of the current best individual. For further details on these important factors, the interested reader is referred to Goldberg (2002).

2.4 Summary and Conclusions

This chapter introduced the three major problem types addressed in this book: (1) optimization problems, (2) classification problems, and (3) reinforcement learning problems. Optimization problems require effective search techniques to find the best solution to the problem at hand. Classification problems require a proper structural partition into different problem classes. RL problems additionally require reward backpropagation.

RL techniques are methods that solve MDP problems online applying dynamic programming techniques in the form of temporal difference learning to estimate discounted future reward. The behavioral policy is optimized according to the estimated reward values. In the simplest case, RL techniques use tables to represent the expected cumulative discounted reward with respect to a state or a state-action tuple.

The most prominent RL technique is Q-learning, which is able to learn an optimal behavioral policy online without learning the underlying state transition function in the problem. Q-learning learns off-policy, meaning that it does not need to pursue the current optimal policy in order to learn the optimal policy.

Genetic algorithms (GAs) are optimization techniques derived from the idea of Darwinian evolution. GAs combine fitness-based selection with mutation and recombination operators to search for better individuals with respect to the current problem. Individuals usually represent complete solutions to a problem.

Effective building block (BB) processing is mandatory in order to solve problems of bounded difficulty. Effective BB processing was recently successfully accomplished using statistical modeling techniques that estimate the dependency structures in the current population and bias recombination towards propagating the identified dependencies.

Niching techniques are very important when the task is to sustain a subset of equally good solutions or different subsolutions for different subspaces (or niches) in the problem space. Fitness sharing and crowding are the most prominent niching methods in GAs.

Goldberg's *facetwise analysis* approach to GA theory significantly improved GA understanding and enabled the design of competent GAs. Although the facetwise approach has the drawback that the found models may need to be calibrated to the problem at hand, the advantages of the approach are invaluable for the analysis and design of highly interactive systems. First, crude models of system behavior are derivable cheaply. Second, analysis is more effective and more general since it is adaptable to the actual problem at hand and focuses only on most relevant problem characteristics. Finally, the approach enables more effective system design and system improvement due to the consequently identifiable rather independent aspects of problem difficulty.

The following chapters investigate how RL and GA techniques are combined in LCSs to solve classification problems and RL problems effectively. Similar to the facetwise decomposition of GA theory and design, we propose a facetwise approach to LCS theory and design in the next chapter. We then pursue the facetwise approach to analyze the XCS classifier system qualitatively and quantitatively. The analysis also leads to the design of improved XCS learning mechanisms and to the proposition of more advanced LCS-based learning architectures.

3

Simple Learning Classifier Systems

Learning Classifier Systems (LCSs) (Holland, 1976; Booker, Goldberg, & Holland, 1989) are rule-based evolutionary learning systems. A basic LCS consists of (1) a set of rules, that is, a *population of classifiers*, (2) a rule evaluation mechanism, which usually is realized by adapted reinforcement learning (RL) (Kaelbling, Littman, & Moore, 1996; Sutton & Barto, 1998) techniques, and (3) a rule evolution mechanism, which is usually implemented by a genetic algorithm (GA) (Holland, 1975). The classifier population codes the current knowledge of the LCS. The evaluation mechanism estimates and propagates rule utility. Based on the estimated utilities, the evolutionary mechanism generates offspring classifiers and deletes less useful classifiers.

LCSs can be distinguished between *online learning* LCSs and *offline learning* LCSs. Moreover, they can be distinguished between LCSs that evolve a single solution, often referred to as Michigan-style LCSs, and LCSs that evolve a set of solutions, often referred to as Pittsburgh-style learning classifier systems (DeJong, Spears, & Gordon, 1993; Llorà & Garrell, 2001b). These systems are usually applied in offline learning scenarios only (also referred to as batch learning).

We analyze LCSs in a modular, facetwise way. That is, different facets of successful LCS learning are analyzed separately and then possible interactions between the facets are considered. The facetwise analysis focuses on appropriate identification, propagation, sustenance, and search of a complete and accurate problem solution. While most of the work focuses on one particular (online learning, Michigan-style) LCS, that is, the accuracy-based learning classifier system XCS (Wilson, 1995), the basic analysis and comparisons as well as the drawn conclusions should readily carry over to other types of LCSs neither restricted to online-learning LCSs nor to Michigan-style LCSs.

This chapter first gives a general introduction to a simple LCS in tutorial form, assuming knowledge about both the basic functioning of a GA as well as basic RL principles. An illustrative example provides more details on basic LCSs. Section 3.3 introduces our facetwise theory approach. Summary and conclusions wrap up the most important lessons of this chapter.

3.1 Learning Architecture

LCSs have a rather simple, but interactive, learning structure combining the strengths of GAs in search, pattern recognition, pattern propagation, and innovation with the value estimation capabilities of RL. The result is a learning system that generates online a generalized state-action value representation. Depending on the complexity of the problem and the number of different states, the generalization capability is able to save space as well as time to learn an optimal behavioral policy. Similarly, in a classification problem scenario, LCSs may be able to detect distributed dependencies in data focusing on the most relevant ones. Conveniently, the dependencies are usually directly reflected in the emerging rules, allowing not only statistical datamining but also more qualitatively oriented datamining.

The basic interaction of the three major components of a learning classifier system and the environment is illustrated in Figure 3.1. While the RL component controls the interaction with the environment, the evolutionary component evolves the problem representation, that is, classifier condition and action parts. Thus, the learning mechanism interacts not only with the environment but also within itself. The evolutionary component relies on appropriate evaluation measures from the RL component and, vice versa, the RL component relies on appropriate classifier structure generated by the GA component to be able to estimate future reinforcement accurately. The interaction between the two components is the key to LCS success. However, proper interaction alone does not assure success. This will become particularly evident in our later analyses. The following paragraphs provide a more concrete definition of a simple learning classifier system LCS1.

3.1.1 Knowledge Representation

To be more concrete, we define a learning classifier system LCS1. LCS1 consists of a population of maximum size N of classifiers. Each classifier cl consists of a condition part $cl.C$, an action part $cl.A$ and a reward prediction value $cl.R$ (using the dot notation to refer to parts of a classifier). Classifier cl predicts reward $cl.R \in \Re$ given its condition $cl.C$ is satisfied, and given further that action $cl.A \in \mathcal{A}$ is executed.

Depending on the representation of the problem space $\mathcal{S}$ (e.g. binary, nominal, real...), conditions may be defined in various ways from simple exact values, over value ranges, to more complex, kernel-based conditions such as radial basis functions. Each classifier condition defines a problem subspace. The population of classifiers as a whole usually covers the complete problem space. Classifiers with non-overlapping conditions (specifying completely different subspaces) are independent with respect to the representation and may be considered as implicitly connected by an **or** operator. Overlapping classifiers compete for activity and reward.

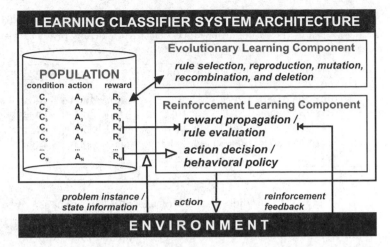

Fig. 3.1. The major components in a learning classifier system are the knowledge base, which is represented by a set of rules (that is, a population of classifiers), the evolutionary component, which evolves classifier structure based on their reward estimation values, and the RL component, which evaluates rules and decides on actions (or classifications).

Let's consider the binary case corresponding to our definition of a Boolean function problem (Chapter 2), as well as our example of Maze 1 (Figure 2.3 on page 17). The binary input string $S = \{0, 1\}^l$ is matched with the conditions that specify the attributes it requires to be correctly set. Traditionally, in its simplest form a condition is represented by the ternary alphabet $C \in \{0, 1, \#\}^l$ where the *don't care* symbol $\#$ matches both zero and one.[1] If the condition part is satisfied by the current problem instance, the classifier is said to *match*. Table 3.1 shows an example of a potential problem instance and all conditions that match this problem instance.

Introducing a little more notation, a classifier *cl* may be said to have a certain *specificity* $\sigma(cl)$. In the binary case, we may define specificity as the ratio of the number of specified attributes to the overall number of attributes. For example, given a problem of length l and a classifier with k specialized (not don't care) attributes, the classifier has a specificity of $\frac{k}{l}$. Similar definitions may be used for other problem domains and other condition representations. Essentially, specificity is a measure that characterizes how much of the problem space a classifier covers. A specificity of one means that only one possible problem instance is covered whereas a specificity of zero means that all problem instances (the whole problem space) is covered. Thus, the

---

[1] Note that the hash symbol might not be expressed explicitly representing a condition part by a set of position-value tuples corresponding to the attributes in the traditional representation that are set to zero or one. This representation has significant computational advantages when the rules only specify a few attributes.

Table 3.1. All classifier conditions whose specified attributes are identical to the corresponding values in the problem instance match the current problem instance. The more general a condition part, the more problem instances it matches.

instance	matching conditions	matching problem instances				condition
1001	1001				1001	1001
	100# 10#1 1#01 #001			1001	1000	100#
	10## 1#0# #00#		1011 1010 1001	1000	10##
	1##1 #0#1 ##01	1111 1101	...	0011	0001	###1
	###1 ##0# #0## 1###	1111 1110	...	0001	0000	####
	####					

larger the specificity of a classifier, the less of the problem space is covered by the classifier. Specificity is an important measure in LCSs, useful for deriving and quantifying evolutionary pressures, problem bounds, and parameter values. Subsequent chapters derive several problem bounds and performance measures based on specificity.

3.1.2 Reinforcement Learning Component

Given a current problem instance, an LCS forms a *match set* $[M]$ of all classifiers in the population $[P]$ whose conditions are satisfied by the current problem instance. The match set reflects the knowledge about the current state (given the current problem instance). An LCS uses the match set $[M]$ to decide on an action.

The action decision is made by the behavioral policy π controlled by the RL component. In the simple case, we can use an adapted ϵ-greedy action selection mechanism. The predicted action value may be decided upon by averaging over the reward predictions of all matching classifiers for each action. The consequent behavioral policy may be written as

$$\pi_{LCS1}(s) = \begin{cases} \arg\ \max_a \frac{\sum_{\{cl\in[M]|cl.A=a\}} cl.R}{|\{cl\in[M]|cl.A=a\}|} & \text{with prob. } 1-\epsilon \\ \text{rand}(a) & \text{otherwise} \end{cases}, \quad (3.1)$$

where s denotes the current problem instance, a the chosen action and $\{cl \in [M]|cl.A = a\}$ the set of all classifiers in the match set $[M]$ whose action part specifies action a. As a result of the action decision $a' = \pi_{LCS1}(s)$ a corresponding action set $[A]$ is formed that consists of all classifiers in the current match set $[M]$ that specify action a' ($[A] = \{cl \in [M]|cl.A = a'\}$).

After the reception of the resulting immediate reward r and the next problem instance s_{+1} yielding match set $[M]_{+1}$, all classifier reward predictions in $[A]$ are updated using the adapted Q-learning equation:

$$R \leftarrow R + \beta(r + \gamma\max_a \frac{\sum_{\{cl\in[M]_{+1}|cl.A=a\}} cl.R}{|\{cl \in [M]_{+1}|cl.A = a\}|} - R), \quad (3.2)$$

estimating the maximum expected discounted future reward by the average of all participating classifiers. The `max` operation indicates the relation to Q-learning. The difference is that in Q-learning only one value is used to estimate a Q-value, whereas in LCS1 a set of classifiers together estimates the resulting Q-value. If all conditions were completely specific, LCS1 would do Q-learning, predicting each Q-value by the means of a single, fully specific classifier.

3.1.3 Evolutionary Component

At this point, we know how reward prediction values are updated and how they are propagated in an LCS. What remains to be addressed is how the underlying conditional structure evolves. Two components are responsible for classifier structure generation and evolution: (1) a covering mechanism, and (2) a GA.

The covering mechanism is mostly applied early in a run to ensure that all problem instances are covered by at least one rule. Given a problem instance, a rule may be generated that sets the value of each attribute with probability $(1 - P_\#)$ to the current value and to a don't care symbol otherwise. Note that covering may be mainly avoided by initializing sufficiently general classifiers. Particularly, if adding classifiers for all possible actions with completely general conditions (all don't care symbols) to the population, covering will not be necessary because the completely general classifiers always match. In this case, the GA needs to fully take care of structure evolution starting from completely (over-)general classifiers.

In its simplest form, we use a steady-state GA (similar to an $(N + 2)$ evolution strategy mechanism (Rechenberg, 1973; Bäck & Schwefel, 1995)). That is, in each learning iteration, the evolutionary component selects two offspring classifiers using, for example, proportionate selection based on the reward predictions R. The selected two classifiers are reproduced, mutated, and recombined yielding two offspring classifiers. For example, mutation can change a condition attribute, with a certain probability μ, to one of the other possible values. Additionally, the action part may be mutated with probability μ. Recombination combines the condition parts with a probability χ applying, for example, uniform crossover. The two offspring classifiers replace two other classifiers, which can be selected using proportionate selection on the inverse of their fitness (e.g. $\frac{1}{1+R}$).

In the case of such a simple GA mechanism, the GA searches in the syntactic (genotypic) local neighborhood of the current population. Selection is biased towards selecting higher reward offspring, consequently propagating classifier structures that predict high reward on average and deleting classifiers that expect low reward on average.

In combination with the RL component, the GA should evolve structures that receive high reward on average. Unfortunately, this is not enough to ensure successful learning. Section 3.3 introduces a general theory of learning in LCSs that reveals the drawbacks of this simple LCS system.

3.2 Simple LCS at Work

Let's do a hypothetical run with our simple LCS1 on the Maze 1 problem (see Figure 2.3 on page 17). Perceptions are coded starting north and coding clockwise indicating an obstacle by 1 and a free position by 0. The money position is perceived as a free position. For example, consider the population shown in Figure 3.2 (generated by hand). Classifier 1 is a classifier that identifies a move to the north whenever there is no obstacle to the east, whereas Classifier 2 considers a move to the north whenever there is no obstacle on the west side. The shown reward values reflect the expected reward received if all situations were equally likely and the correct Q-values were propagated (effectively an approximation of the actual values).

In the shown iteration, the provided problem instance 01011101 indicates that there is a free space north, east, and west (as a result of residing in the position just south of the money position). The problem instance triggers the formation of a match set, as indicated in Figure 3.2. In the example, the classifiers shown in the match set predict a reward of 950.75 for action ↑ and 900 for action ↓. When action ↑ is executed, the money position is reached, a reward r of 1000 is received, and the reward predictions of all classifiers in the current action set are updated, applying Equation 3.2 (shown are updates using learning rate $\beta = 0.2$). It can be seen how the reward estimation values of all classifiers increase towards 1000.

Finally, a GA is applied that selects two classifiers from the population, reproduces, mutates, and recombines them, and replaces two existing classifiers by the new classifiers. For example, the GA may select classifiers three and six, reproducing them, mutating them to e.g. $3'=$(0####1###,↑) and $6'=$(0101#101,↑), recombining them using one-point crossover to e.g. $3^*=$(0####1101,↑) and $6^*=$(0101####,↑), and finally reinserting 3^* and 6^* into the population, replacing two other classifiers (e.g. the lower reward classifiers two and five).

We can see that the evolutionary process propagates classifiers that specify how to reach the rewarding position. Due to the bias of reproducing classifiers that predict higher reward, on average, higher-reward classifiers will be reproduced more often and will be deleted less often. Mutation and crossover serve as the randomized search operators that are looking for better solutions in the (syntactically) local neighborhood of the reproduced classifiers.

Our example already exhibits several fundamental challenges for simple learning classifier systems: the problem of *strong overgenerals* (Kovacs, 2000), investigated in detail elsewhere (Kovacs, 2003), the problem of generalization, and the problem of local vs. global competition. These issues are the subject of the following section, in which we develop a facetwise theory approach for LCS analysis and design.

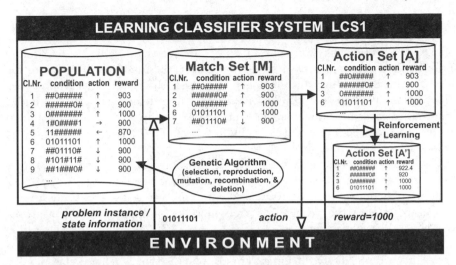

Fig. 3.2. In a typical learning iteration, an LCS receives a current problem instance consequently forming the match set $[M]$. Next, an action is chosen (in this case action ↑) and executed in the environment. Since the resulting reward equals one thousand, the reward estimates are increased (shown is an increase using learning rate $\beta = 0.2$). Finally, the GA may be applied on the updated population.

3.3 Towards a Facetwise LCS Theory

The introduction of LCS1 should have clarified several important properties of the general LCS learning architecture, such as the online learning property and the interaction of rule evaluation (using reinforcement methods) and rule structure learning (using evolutionary computation methods). However, it remains to be understood if and how the interactions can be assured to learn a complete problem solution.

It is clear that on average classifiers which receive a higher reward will be selected for reproduction more often and will be deleted less often. Thus, the GA mechanism propagates their structure by searching in the local neighborhood of these structures. Can we make specific learning projections? How general will the evolved solution be? How big can the problem be, given a certain population size? How distributed will the final population be?

This section addresses these issues and develops a facetwise theory for LCSs to answer them in a modular way. The section first gives a general outlook of which solution LCSs evolve and *how* this may be accomplished. Next, a theory approach, which addresses *when* this may be accomplished, is proposed.

3.3.1 Problem Solutions and Fitness Guidance

The above sections showed that LCSs are designed to evolve a distributed problem solution represented by a population of classifiers. The condition of each classifier defines the subspace for which the classifier is responsible. The action specifies the proposed solution in the defined subspace. Finally, the reward measure estimates the consequent payoff of the chosen action. It is desired that the action with the highest estimated payoff equals the best action in the current problem state.

To successfully evolve such a distributed problem solution we need to prevent overgeneralization and we rely on sufficient fitness guidance (that is, a fitness gradient) towards better classifiers. The two issues are discussed below starting with the problem of overgenerality and the problem of strong overgenerals in particular.

The problem of *strong overgenerals* (Kovacs, 2001) concerns a particular generality vs. specificity dilemma. The problem is that a general classifier cl_g (one whose conditions are satisfied in many problem instances or states) may have a higher reward prediction value, on average, than a more specialized classifier cl_s, which may match in a subset of cl_g. Consequently, the GA will propagate cl_g. Additionally, given that the actions are different in cl_g and cl_s, action $cl_g.A$ has preference over action $cl_s.A$. However, action $cl_s.A$ may yield higher reward in situations in which cl_s also matches. Thus, although action $cl_s.A$ would be more appropriate in the described scenario, action $cl_g.A$ will be chosen by the behavioral policy and will be propagated by the GA. Such an overgeneral classifier is called *strong overgeneral* classifier because it has a higher reward estimate than other, more specific classifiers and is consequently stronger with respect to evolutionary reproduction and action selection. The following example helps to clarify the problem.

Considering classifiers 1 and 4 in Figure 3.2, it can be seen that both classifiers match in the left-most position of Maze 1 (Figure 2.3 on page 17), which is perceived as 11011111. The best action in this position is certainly to move to the east, which yields a discounted reward of 900 (as correctly predicted by Classifier 4). However, Classifier 1 predicts a slightly higher reward for action ↑, since its reward reflects the average reward encountered when executing action ↑ in its matching states A, B, C, and D. The values would be exact if the classifier was updated sufficiently often, the learning rate was sufficiently small, and all states were visited equally often, and action north was executed equally often in each of the states. Thus, the incorrect action ↑ may be executed (dependent on the other classifiers in the population) although the action → would be correct.

The basic problem shows that good classifiers cannot be distinguished from bad classifiers as easily as initially thought. In effect, it needs to be questioned if the fitness approach—deriving fitness directly from the absolute reward received—is appropriate, or rather, in which problems a direct reward-based fitness approach is appropriate. Kovacs (Kovacs, 2001; Kovacs, 2003)

analyzes this problem in detail. The most important result is that strong
overgenerals can occur in any problem in which more than two reward values
may be perceived (and thus essentially in all but the most trivial multistep
problems). The severeness of this problem consequently demands that the
fitness approach itself should be changed.

Solutions to this problem are the *accuracy-based* fitness approach in XCS
(Wilson, 1995), investigated in detail in later chapters, and the application of
fitness sharing techniques, as applied in ZCS (Wilson, 1994). In the former
case, local fitness is modified requiring explicitly that the reward prediction
of a classifier is accurate (that is, it has low variance). In the latter case,
reward competition in the problem subspaces (defined by the current problem
instance) can cause the extinction of strong overgeneral classifiers.

The problem of strong overgenerals illustrates how important it is to exactly define (1) the structure of the problem addressed (to know which challenges are expected) and (2) the objective of the learning system (to know how
the system may "misbehave"). Our strength-based system LCS1, for example,
works sufficiently well (with respect to strong overgenerals) in all classification problems since only two reward values are received and reward is not
propagated. However, other challenges may have to be faced as investigated
below.

Once we can assure that classifiers in the optimal solution will have the
highest fitness values, we need to ensure that fitness itself *guides* the learning process towards these optimal classifiers. It should be acknowledged that
overgeneral classifiers should have lower fitness values by definition of the optimal solution. To what degree fitness guides towards higher fitness values
depends on the fitness definition and problem properties. Later chapters address this problem in detail with respect to the XCS classifier system and
typical problem properties.

3.3.2 General Problem Solutions

While the problem of strong overgenerals is concerned with a particular phenomenon resulting from the interaction of a classifier structure, reinforcement component, and evolutionary computation component, the problem also
points to a much more fundamental problem: the problem of generalization.
When we introduced LCSs above, we claimed that they can be characterized as online generalizing RL systems. And in fact, as the classifier structure
suggests, rules are often matching in several potential problem instances or
states. However, until now it was not addressed at all why and how the evolutionary component may propagate more general classifiers instead of more
specific ones.

Considering again our Maze 1 (Figure 2.3 on page 17) and the exemplar
population shown in Figure 3.2, classifiers 6 and 3 actually contain the same
amount of information: Both classifiers only match in maze position C and
both classifiers predict that a move to the north yields a reward of 1000.

Clearly, Classifier 6 is syntactically much more specialized. The general concept of Classifier 3, which only requires a free space to the north and does not care about any other position, appears more appealing and might be the best concept in the addressed environment. In general, the aim is to stay as general as possible, identifying the minimal set of attributes necessary to predict a Q-value correctly.

On the other hand, consider classifiers 7 and 8 in Figure 3.2. Both classifiers predict that moving south results in a reward of 900. Both classifiers are syntactically equally specific, that is, both have an order of five (five specified attributes). However, Classifier 7 is semantically more general than Classifier 8 because it matches in more states than Classifier 8 (all three states below the money vs. only the states south and southeast of the money). Classifier 9 is semantically as general as Classifier 7 but it is syntactically more general. Later, we will see that XCS biases its learning towards syntactic and semantic generality using different mechanisms.

In the general case, the quest for generality leads us to a multi-objective problem in which the objectives are to learn the underlying problem as accurately as possible and to represent the solution with the most general classifiers and the least number of classifiers possible. This problem is addressed explicitly elsewhere (Llorà & Goldberg, 2003; Llorà, Goldberg, Traus, & Bernadó, 2003), in which a Pareto-front of high fitness, high generality classifiers is propagated. Other approaches, including the mechanism in XCS, apply a somewhat constant generalization pressure that is overruled by the fitness pressure if higher fitness is still achievable. Yet another generalization approach, recently proposed by Bull (2003), was applied in the ZCS system. In this case, reproduction causes fitness deduction. The lost fitness can only be regained by reapplication. More general classifiers will be reapplied faster and thus do not suffer as much from the reproduction penalty and eventually take over the population.

In Chapter 5, the effects of different generalization mechanisms are analyzed in more detail.

3.3.3 Growth of Promising Subsolutions

Once we know which solution we intend to evolve with our LCS system, how fitness may guide us to the solution, and how the solution will tend to be general, we need to ensure that our intentions can be put into practice. Thus, we need to ensure the growth of higher fitness classifiers.

To do this, it is necessary to ensure that classifiers with higher fitness are available in the population. Once we can assure that better classifiers are present, we need to assure that the RL component and the genetic component can interact successfully to reproduce higher fitness classifiers. Therefore it is necessary to assure that the RL component has enough *evaluation time* to detect higher fitness classifiers reliably. Moreover, the genetic component needs to reproduce and thus propagate those better classifiers before they tend to be deleted.

The first aspect is related to the *BB supply* issue in GAs. However, due to the distributed problem representation, a more diverse supply may need to be ensured, and the definition of supply differs. In essence, the initial population needs to be general enough to cover the whole problem space, but it also needs to be diverse enough to have better solutions available for identification and propagation. The diversification and specialization effects of mutation may support the supply issue. These ideas become much more concrete when investigating the XCS classifier system in Chapter 6.

Note that supply is not only relevant in the beginning of a run, but it is actually relevant at all stages of the learning progress, continuously requiring the supply or generation of better offspring classifiers. However, the issue is most relevant in the beginning due to the fact that later in the run, the currently found distributed problem solution usually significantly restricts the search space to the immediate surrounding of these solutions. In the beginning of a run, the whole search space is the surrounding and any randomized search operator, such as mutation, can be expected—dependent on the problem—to have a hard time to find better classifiers by chance.

Once better classifiers are available, we need to ensure that they are identified. Since the RL component requires some time to identify better classifiers (iteratively updating the reward estimates), better classifiers need to have a sufficiently long survival time. Thus, offspring classifiers need to undergo several evaluations before they are deleted.

Finally, if better classifiers are available and the RL component has enough time to identify them, it is necessary to ensure that the genetic component propagates them. Thus, the survival time also needs to be long enough to ensure the reproduction of better classifiers. Additionally, genetic search operators may need additional time to effectively detect important problem substructures and subspaces. Due to the potentially unequally distributed problem complexity in problem space (see, for example, the problem in Figure 3.6 on page 46), different time may need to be available for different problem subspaces. Chapter 6 investigates these issues in detail with respect to the XCS classifier system.

3.3.4 Neighborhood Search

Once we can assure that higher fitness classifiers undergo reproduction and thus grow in the population, we need to implement effective neighborhood search in order to detect even better problem solutions. These problem solutions can be expected to lie in the neighborhood of the currently best subsolution or in further partitions of the current subsolution subspace, defined by the classifier conditions. The neighborhood search especially is very problem dependent and thus it is impossible to define generally optimal search operators. We now first look at simple mutation and crossover and their impact on genetic search. Next, we discuss the issue of local vs. global search bias in somewhat more detail.

Mutation

Mutation generally searches in the syntactic neighborhood of a selected classifier. A simple mutation operator changes some attributes in the condition part of a classifier as well as the class of the classifier. If the class is changed, the new classifier basically considers the possibility that the relevant attributes for one class might also be appropriate for a reward prediction in another class. This might be helpful especially in multistep problems, where classifiers often develop conditions that identify sets of states that are equally distant from reward.

Mutation of the condition part can have three types of effects that may apply in combination if several attributes of one condition part are mutated: (1) generalization, (2) specialization, (3) knowledge transfer. Considering the ternary alphabet $C \in \{0, 1, \#\}^l$ and given an attribute with value 0, mutation may change the attribute to $\#$. In this case, the classifier is generalized (its specificity decreases) since its condition covers a larger problem subspace (in the binary case, double the space). On the other hand, if the attribute actually was a don't care symbol before and it is mutated to 0 or 1, the classifier is specialized (its specificity is increased) covering a smaller portion of the problem space (in the binary case, half of the space). Finally, a specified attribute (e.g. 0) may be changed to another specific value (e.g. 1), effectively transferring the subspace structure of the rest of the classifier condition to another part of the search space.

Figure 3.3 illustrates the three mutation cases. Given the parental classifier condition 11#11#, cases (a) and (b) show the potential cases for specialization, that is 11011#, 11111# and 11#110, 11#111, respectively. Case (c) shows knowledge transfer when a specialized attribute is changed to the other value. In our example, the classifier condition may change to 11#10#, 11#01#, 10#11#, or 01#11#, effectively moving the hyperrectangle to other subspaces in the problem space. Cases (d), (e), and (f) show how generalization by mutation may change the condition structure. Note that in each case, the original hyperrectangle is maintained and another hyperrectangle with a similar structure is added. The shown cases cover all possible mutation cases of one attribute in the parental classifier. Depending on the mutation probability μ, additional mutations are exponentially less probable, but may result in an extended neighborhood search.

In effect, mutation searches in the general/specific neighborhood of the current solution and it transfers structure from one subspace to another (near-by) subspace. The effectiveness of mutation consequently depends on the complexity distribution over the search space. If syntactic neighborhoods are structured similarly, mutation can be very effective. If there are strong differences between syntactic neighborhoods, mutation can be quite ineffective.

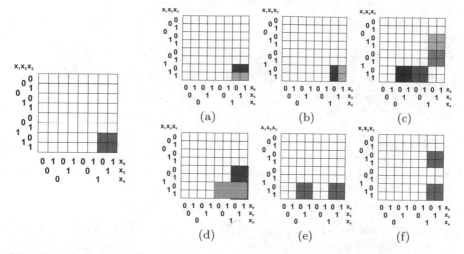

Fig. 3.3. Mutation in action: The parental classifier condition on the left-hand side can be either specialized by one mutation (a,b), projected into a neighboring subspace (c), or generalized including a neighboring subspace (d,e,f). Different grays represent different classifier conditions.

Regardless of the problem search effect, mutation is a general diversification operator that causes the evolutionary search procedure to search in the local (syntactic) neighborhood of currently promising solutions. Doing this, mutation tends to generate an equal number of each available value for an attribute. In the ternary alphabet, mutation consequently pushes the population towards an equal amount of zeros, ones, and #-symbols in classifier conditions. With respect to specificity, mutation pushes towards a 2:1 specific:general distribution. Thus, in the usual case when starting with a fairly general classifier population, mutation has a specialization effect. How mutation influences specificity in XCS and how the specificity influence interacts with other search operators in the system is analyzed in Chapter 5.

Crossover

The nature of crossover strongly differs from mutation in that crossover does not search in the local neighborhood of one classifier, but combines classifier structures. The results are offspring classifiers that specify substructures of the parental classifiers. In contrast to mutation, simple crossover does not affect overall specificity, since the combined specificity of the two offspring classifiers equals the combined specificity of the parents. Thus, although the specificity of individual offspring classifiers might differ from the parental specificity, average specificity is not affected.

For example, consider the overlapping classifier conditions 11#### and 1##00# shown on the left-hand side of Figure 3.4. The maximal space crossover

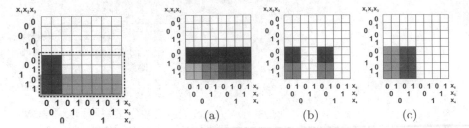

Fig. 3.4. Crossover of two overlapping parental conditions searches in the subspace indicated by the outer dashed box (left-hand side). More specialized offspring conditions (shown in brighter gray) are included in the more general offspring conditions (shown in increasingly darker gray).

searches in is restricted to the maximal general offspring that can be generated from the two parental classifiers, which is 1##### in our example. Other offspring structures are possible, which are progressively closer to the parental structures as indicated in Figure 3.4 (a), showing progressively more specialized classifier conditions as well as in (b) and (c) showing the four other possible offspring cases. It can be seen that crossover consequently searches in the maximal problem subspace defined by the two classifier conditions. Structure of the two classifiers is exchanged and projected onto other subspaces inside the maximal subspace.

If the parental classifiers are non-overlapping, crossover searches in the maximum subspace, which is defined by the non-overlapping parts of the two subspaces. In the example shown in Figure 3.5, the parental classifiers 1##00# and 00#### are non-overlapping and the maximum subspaces are characterized by either the upper half of the search space (0#####) or the lower half of the search space (1#####). Note that in the case of non-overlapping classifiers, the structural exchange may or may not be fruitful and strongly depends on the underlying problem structure. If structure is similar throughout the whole search space, then crossover may be beneficial. However, if structure differs in different search subspaces, crossover can be expected to be mainly disruptive, when non-overlapping conditions are recombined.

In general, crossover recombines previously successful substructures, transferring those substructures to other, nearby problem subspaces. Depending on the complexity and uniformity of problem spaces, crossover may be more or less effective. Also, it can be expected that the recombination of classifiers that cover structurally related problem spaces will be more effective than the recombination of unrelated classifiers. Thus, a good restriction of classifier recombination is expected to result in a more effective genetic search.

As in GAs, the issue of building blocks (BBs) comes into mind. In simple LCSs, BBs are complexity units, which define a subspace that yields high reward on average. The identification and effective propagation of BBs, consequently, should result in another type of more effective search within LCSs.

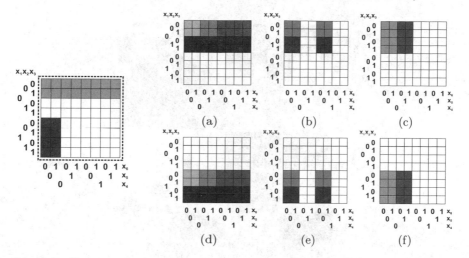

Fig. 3.5. Given two non-overlapping parental conditions (left-hand side), crossover explores structure in either of the resulting maximal general, non-overlapping subspaces. Figures a,b,c,d,e,f show condition subspaces that can be generated by crossover. Brighter gray subspaces are included in darker subspaces but also form a condition subspace on their own.

Considering such search biases, however, it needs to be kept in mind that we are searching for a distributed solution representation in which different subsolutions consist of different BB structures.

Local vs. Global Search Bias

Due to the distributed problem solution representation, LCSs face another challenge in comparison to standard GAs. Although the incoming problem instances must be assumed to be structurally and semantically related, the currently evolved problem subsolutions may be unequally advanced, and good subsolutions may be highly unequally complex, depending on the problem subspace they apply in. For example, in some regions of the problem space, a very low specificity might be sufficient to predict reward correctly whereas in other regions further specificity might be necessary.

Figure 3.6 illustrates a problem space in which some subspaces are highly complex (subspace 00******), in that the identification of the problem class (black or white) requires several further feature specifications. On the other hand, the rest of the problem space is fairly simple (subspaces 01******, 10******, and 11******) in that classes can be distinguished easily by the specification of only one or two additional features (01**1***→1, 10**0***→1, and 110**1**→1 and 111**0**→1).

Similar differences in complexity can be found in RL problems. For example, in the simple Maze problem in Figure 2.3 on page 17 and the exemplar

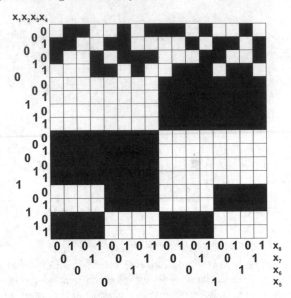

Fig. 3.6. Problem subspaces may vary in complexity dependent on the problem structure. Complexity-dependent niching as well as suitable subspace-dependent search mechanisms are mandatory to enable the effective evolution and sustenance of a complete problem representation.

population shown in Figure 3.2 on page 37, we can see that a classifier of specificity 1/8 (order one—specifying the empty position to the north) suffices to predict a reward of 1000 correctly when going north (Classifier 3). On the other hand, in order to predict a 900 reward correctly when heading south, at least two positions need to be specified (see Classifier 9). Thus, necessary specificities as well as necessary specified positions might differ depending on the problem instance. In effect, global selection and recombination might not be appropriate since different classifiers might represent solutions to completely different subspaces. Thus, a balanced search combining local and global knowledge can be expected to be most effective.

In particular, it is desired to only recombine those classifiers that are compatible in the sense that they address related problems as specified by their condition parts. In the simple LCS, selection, recombination, and crossover are usually applied globally in that two potentially unrelated classifiers are selected from the overall population. The consequent recombination is then likely to be ineffective if solution structure varies over the problem space. We will see that XCS circumvents this problem by reproducing classifiers in action sets. However, if the problem has a much more global structure, further bias towards global selection may result in additional learning advantages. We address this problem in further detail in Chapter 7.

3.3.5 Solution Sustenance

As seen above, a simple GA selects individuals from the whole population, mutates, and potentially recombines them. Since a GA is usually designed to optimize a problem, it usually searches for one globally optimal solution. The sustenance of different solutions is important only early in the run in this case. Sustenance is usually assured through initial population diversity and supply as well as by balancing the focusing effects of selection with the diversification effects of recombination and/or mutation operators.

A different problem arises if the goal is to evolve not only the best solution but a distributed set of solutions. Since LCSs are designed to generate the best solution for every potential problem instance, the population in an LCS needs to evolve optimal solutions for all potential problems in the problem space. Each problem instance represents a new (sub-)problem (defined by the problem instance) that might be related to the other (sub-)problems but represents a new problem in a common problem space.

Due to the necessity of a distributed representation, niching methods (Goldberg & Richardson, 1987; Horn, 1993; Horn, Goldberg, & Deb, 1994; Mahfoud, 1992) are even more important in LCSs than they are in standard GAs. Since LCSs need to evolve and maintain a representation that is able to generate a solution for every possible problem instance, it needs to be assured that the whole problem space is covered by the distributed solution representation. Several niching methods are applicable and different LCSs use different techniques to assure the sustenance of a complete problem solution. Later chapters address the niching issue in further detail.

3.3.6 Additional Multistep Challenges

The above issues mainly targeted problems in which immediate reward indicates the appropriateness of an action. Also, problem instances were thought to be independent of each other. In multistep problems, such as the ones modeled by general MDP or POMDP processes, reward propagation as well as self-controlled problem sampling comes into play.

Evaluation and reproduction time issues need to be reconsidered in this case. Since successive problem instances depend on the executed action, which is chosen by the LCS agent itself, problem instance sampling and problem instance frequencies may become highly skewed. Thus, the time issues with respect to classifier selection and propagation in problem subspaces may need to be reevaluated.

Additionally, the RL and GA component may be influenced by the current action policy and vice versa, the action policy may depend on the RL and GA constraints. For example, exploration may be enhanced in problem subspaces in which no appropriate classifier structure evolved so far. This relates to prioritized sweeping and other biased search algorithms in RL (Moore & Atkeson, 1993; Sutton, 1991).

Besides sampling and learning bias issues, the challenge of reward propagation needs to be faced. Since reinforcement may be strongly delayed depending on the problem and current problem subspace, accurate non-disruptive reward backpropagation needs to be ensured. Thus, competent RL techniques need to be applied. Additionally, due to the rule generalization component, update techniques that weigh the confidence of the predictive contributions of each classifier, similar to error estimation techniques used in neural networks, may be advantageous. These concerns are analyzed in detail in Chapter 10.

3.3.7 Facetwise LCS Theory

The above issues lead to the proposition of the following LCS problem decomposition. To assure that a classifier system evolves an accurate problem solution, the following aspects need to be respected:

1. *Design evolutionary pressures most effectively*:
   Appropriate fitness definition, parameter estimation, and rule generalization.
   a) **Fitness guidance**: Fitness needs to be defined appropriately to guide towards the optimal solution disabling strong overgenerals.
   b) **Parameter estimation**: Classifier parameters need to be initialized and estimated most effectively to avoid fitness disruption in young and unexperienced classifiers as well as to identify better classifiers as fast as possible.
   c) **Adequate generalization**: Classifier generalization needs to push towards a maximally accurate, maximally general problem solution preventing overgeneralization.
2. *Ensure solution growth and sustenance*:
   Effective population initialization, classifier supply, classifier growth and niche sustenance.
   a) **Population initialization**: Effective classifier initialization needs to ensure sufficient classifier evaluation and GA application time.
   b) **Schema supply**: Minimal order schema representatives need to be available.
   c) **Schema growth**: Schema representatives need to be identified and reproduced before deletion is expected.
   d) **Solution sustenance**: Niching techniques need to ensure the sustenance of a complete problem solution.
3. *Enable effective solution search*:
   Effective mutation, recombination, and structural biases.
   a) **Effective mutation**: Mutation needs to search the neighborhoods of current subsolutions effectively ensuring diversity and supply.
   b) **Effective recombination**: Recombination needs to combine building block structures efficiently.

    c) **Local vs. global structure**: Search operators need to detect and exploit global structural similarities but also differences in different local problem solution subspaces.

4. *Consider additional challenges in multistep problems*: Behavioral policies, sampling biases, and reward propagation.

    a) **Effective behavior**: A suitable behavioral policy needs to be installed to ensure appropriate environmental exploration and knowledge exploitation.

    b) **Problem sampling reconsiderations**: Subproblem occurrence frequencies might be skewed due to environmental properties and the chosen behavioral policy. Thus, evaluation and reproduction times need to be reevaluated and might be synchronized with currently chosen behavior and encountered environmental properties.

    c) **Reward propagation**: Accurate reward propagation needs to be ensured to allow accurate classifier evaluation.

The next chapters focus on this LCS problem decomposition, investigating how the accuracy-based XCS classifier system faces and solves these problem aspects. Along the way, we also show important improvements in the XCS system in the light of several of the theory facets.

3.4 Summary and Conclusions

In this chapter we introduced the general structure of an LCS. We saw that LCSs are suitable to learn classification problems as well as RL problems. Additionally, we saw that LCSs are online learning systems that learn from one problem instance at a time, potentially interacting with a real environment or problem.

LCSs learn a distributed problem solution represented by a population of classifiers (that is, a set of rules). Each classifier specifies a condition, which identifies the problem subspace in which it is applicable, an action or classification, and a reward prediction, which characterizes the suitability of that action given the current situation. Although we only considered conditions in the binary problem space, applications in other problem spaces including nominal and real-valued inputs are possible. Chapter 9 investigates the performance of the accuracy-based classifier system XCS in such problem spaces.

Classifiers in an LCS population are *evaluated* by RL mechanisms. Classifier structure is *evolved* by a steady-state GA. Thus, while the RL component is responsible for the identification of better classifiers, the GA component is responsible for the propagation of these better classifiers. The consequent strong interdependence of the two learning components needs to be considered when creating an LCS.

Dependent on the classifier condition structure, mutation and crossover have slightly different effects in comparison to search in a simple GA. Mutation changes the condition structure searching for other subsolutions in the neighborhood of the parental condition. However, mutation does not only cause a diversification pressure, it can also have a direct effect on the *specificity* of the condition of a classifier. Crossover, on the other hand, does not affect combined classifier specificity but recombines problem substructures. As in GAs, crossover may be disruptive and BB identification and propagation mechanisms may improve genetic search. In contrast to GAs, though, LCSs search for a distributed problem solution so that crossover operators need to distinguish and balance search influenced by local and global problem structure.

The proposed facetwise LCS theory approach for analysis and design is expected to result in the following advantages: (1) Simple computational models of the investigated LCS system can be found, (2) The found models are generally applicable, (3) The models are easily modifiable to the actual problem and representation at hand, (4) The models provide a deeper and more fundamental problem and system understanding, (5) The investigated system can be improved effectively focusing on the currently most restricting model facets, and (6) More advanced LCS systems can be designed in a more straightforward manner, targeted to effectively solve the problem at hand. The remainder of this book pursues the facetwise analysis approach, which confirms the expected advantages.

4

The XCS Classifier System

The creation of the accuracy-based classifier system XCS (Wilson, 1995) can be considered a milestone in classifier system research. XCS addresses the general LCS challenges in a very distributed fashion. The problem of generalization is approached by niche reproduction in conjunction with panmictic (population-wide) deletion. The problem of strong overgenerals is solved by deriving classifier fitness from the estimated accuracy of reward predictions instead of from the reward predictions themselves. In effect, XCS is designed to not only evolve a representation of the best solution for all possible problem instances but rather to evolve a complete and accurate *payoff map* of all possible solutions for all possible problem instances.

This chapter introduces the XCS classifier system. We provide a concise description of problem representation and all fundamental mechanisms. The algorithmic description found in Appendix B provides an exact description of all mechanisms in XCS facilitating the implementation of the system. After the introduction of XCS, Section 4.2 shows XCS's performance on simple toy problems. Chapters 5 and 6 investigate how and when XCS is able to learn a solution developing a theory of XCS's learning capabilities. Subsequent chapters then evaluate XCS in more challenging problems, analyze learning and scalability, and introduce several improvements to the system.

4.1 System Introduction

As with most LCSs, XCS evolves a set of rules, the so-called *population* of *classifiers*, by the means of a steady-state GA. A classifier usually consists of a condition and an action part. The condition part specifies when the classifier is applicable and the action part specifies which action, or classification, to execute. XCS differs from other LCSs in its GA application and its fitness approach. This section gives a concise introduction to XCS starting with knowledge representation, progressing to learning iteration, and ending with learning evaluation and the genetic learning component. The introduction focuses on binary problem representations. Nominal and real-valued representations are introduced in Chapter 9.

4.1.1 Knowledge Representation

The population of classifiers represents a problem solution in a probabilistic disjunctive normal form where each classifier specifies one conjunctive term in the disjunction. Thus, each classifier can be regarded as an expert in its problem subspace specifying a confidence value about its expertise.

Each classifier consists of five main components and several additional estimates.

1. *Condition part C* specifies when the classifier is applicable
2. *Action part A* specifies the proposed action (or classification or solution)
3. *Reward Prediction R* estimates the average reward received when executing action A given condition C is satisfied
4. *Reward prediction error ε* estimates the mean absolute deviation of R with respect to the actual reward
5. *Fitness F* estimates the scaled, relative accuracy (scaled, inverse error) with respect to other, overlapping classifiers

As in LCS1, in the binary case condition part C is coded by $C \in \{0, 1, \#\}^{L}$ identifying a hyperrectangle in which the classifier is applicable, or *matches*. Action part $A \in \mathcal{A}$ defines one possible action or classification. Reward prediction $R \in \Re$ is iteratively updated resulting in a moving average measure of encountered reward received in the recent problem instances in which condition C matched and action A was executed. Similarly, the reward prediction error estimates the moving average of the absolute error of the reward prediction. Finally, fitness estimates the moving average of the accuracy of the classifier's reward prediction relative to other classifiers that were applicable at the same time.

Each classifier maintains several additional parameters. The *action set size estimate as* estimates the moving average of the action sets it is applied in. It is updated similar to the reward prediction R. The *time stamp ts* specifies the time when last the classifier was part of a GA competition. The *experience counter exp* counts the number of parameter updates the classifier underwent so far. The numerosity *num* specifies the number of identical micro-classifiers, this (macro-)classifier actually represents. In this way, multiple identical classifiers can be represented by one actual classifier in the population speeding up computation (for example, for the matching process).

4.1.2 Learning Interaction

Learning usually starts with an empty population. Alternatively, the population may be initialized generating random classifiers whose condition parts have an average specificity of $1 - P_\#$ (that is, each attribute in the condition part is a don't care symbol with probability $P_\#$ and zero or one otherwise).

Given current problem instance $s \in \mathcal{S}$ at iteration time t, the *match set* $[M]$ is formed, containing all classifiers in $[P]$ whose conditions match s. If some action is not represented in $[M]$, a covering mechanism is applied.[1] Covering creates classifiers that match s (inserting #-symbols, similar to when the population is initialized at random, with a probability of $P_\#$ at each position) and specify unrepresented actions. Given a match set, XCS can estimate the payoff for each possible action forming a *prediction array* $P(\mathcal{A})$,

$$P(A) = \frac{\sum_{cl.A=A \wedge cl \in [M]} cl.R \cdot cl.F}{\sum_{cl.A=A \wedge cl \in [M]} cl.F}. \tag{4.1}$$

We use the dot notation to refer to classifier parameters in XCS to avoid confusion with classifier unrelated parameters. Essentially, $P(A)$ reflects the fitness-weighted average of all reward prediction estimates of the classifiers in $[M]$ that advocate classification A. The prediction array is used to determine the appropriate classification. Several action selection policies (that is, behavioral policies) may be applied. Usually, XCS chooses actions randomly during learning, and it chooses the best action $A_{max} = \texttt{arg max}_A P(A)$ during testing. All classifiers in $[M]$ that specify the chosen action A comprise the action set $[A]$.

After the execution of the chosen action, feedback is received in the form of scalar reward $R \in \Re$, which is used to update classifier parameters. Finally, the next problem instance is received and the next problem iteration begins, increasing the iteration time t by one.

4.1.3 Rule Evaluation

XCS iteratively updates its population of classifiers with respect to the successive problem instances. Parameter updates are usually done in this order: prediction error, prediction, fitness.

In a classification problem, classifier parameters are updated with respect to the immediate feedback R in the current action set $[A]$. In an RL problem, all classifiers in $[A]$ are updated with respect to the immediate reward R plus the estimated discounted future reward as follows:

$$Q = max_{A \in \mathcal{A}} P^{t+1}(A), \tag{4.2}$$

where the $(t+1)$ term refers to the prediction array in the consequent learning iteration $t + 1$.

Reward prediction error ε of each classifier in $[A]$ is updated by:

---

[1] Covering is sometimes controlled by the parameter θ_{mna} that requires that at least θ_{mna} actions are covered. For simplicity, we set θ_{mna} per default to the number of possible actions or classifications $|\mathcal{A}|$ in the current problem.

$$\varepsilon \leftarrow \varepsilon + \beta(|\rho - R| - \varepsilon), \tag{4.3}$$

where $\rho = r$ in classification problems and $\rho = r + \gamma Q$ in RL problems. Parameter $\beta \in [0,1]$ denotes the learning rate influencing accuracy and adaptivity of the moving average reward prediction error. Similar to the learning rate dependence in RL (see Chapter 2), a higher learning rate β results in less history dependence and thus faster adaptivity but also in a higher variance if different reward values are received.

Next, reward prediction R of each classifier in $[A]$ is updated by:

$$R \leftarrow R + \beta(\rho - R), \tag{4.4}$$

with the same notation as in the update of ε. Note how XCS essentially applies a Q-learning update. However, Q-values are not approximated by a tabular entry but by a collection of rules expressed in the prediction array $P(\mathcal{A})$.

The fitness value of each classifier in $[A]$ is updated with respect to its current scaled relative accuracy κ', which is derived from the current reward prediction error ε as follows:

$$\kappa = \begin{cases} 1 & \text{if } \varepsilon < \varepsilon_0 \\ \alpha \left(\frac{\varepsilon}{\varepsilon_0} \right)^{-\nu} & \text{otherwise} \end{cases}, \tag{4.5}$$

$$\kappa' = \frac{\kappa \cdot num}{\sum\limits_{cl \in [A]} cl.\kappa \cdot cl.num}. \tag{4.6}$$

Essentially, κ measures the current absolute accuracy of a classifier using a power function with exponent ν to further prefer low error classifiers. Threshold ε_0 denotes a threshold of maximal error tolerance. That is, classifiers whose error estimate ε drops below threshold ε_0 are considered accurate. The derivation of accuracy κ with respect to ε is illustrated in Figure 4.1. The relative accuracy κ' then reflects the relative accuracy with respect to the other classifiers in the current action set. In effect, each classifier in $[A]$ competes for a limited fitness resource that is distributed dependent on $\kappa \cdot num$.

Finally, fitness estimate F is updated with respect to the current action set relative accuracy κ' as follows:

$$F \leftarrow F + \beta(\kappa' - F). \tag{4.7}$$

In effect, fitness loosely reflects the moving average, set-relative accuracy of a classifier. As before, β controls the sensitivity of the fitness.

Additionally, the action set size estimate as is updated similar to the reward prediction R but with respect to the current action set size $\|[A]\|$:

$$as \leftarrow as + \beta(\|[A]\| - as), \tag{4.8}$$

revealing a similar sensitivity to action set size changes $\|[A]\|$ dependent on learning rate β.

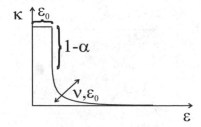

Fig. 4.1. The derivation of accuracy κ from the current reward prediction error ε has an error tolerance of ε_0. Exponent ν controls the degree of the drop off and ε_0 scales the drop off. Parameter α differentiates accurate and inaccurate classifiers further.

Parameters R, ε, and as are updated using the *moyenne adaptative modifiée* (which could be translated as adaptive average-based modification) technique (Venturini, 1994). This technique sets parameter values directly to the average of the so far encountered cases as long as the experience of a classifier is less than $1/\beta$.

Each time the parameters of a classifier are updated, experience counter exp is increased by one. Additionally, if genetic reproduction is applied to classifiers of the current action set, all time stamps ts are set to the current iteration time t.

Using the Widrow-Hoff delta rule, the reward prediction of a classifier approximates the mean reward it encounters in a problem. Thus, the reward prediction of a classifier can be approximated by the following estimate:

$$cl.R \approx \frac{\sum_{\{s|cl.C \text{ matches } s\}} p(s)p(cl.A|s)\rho(s, cl.A)}{\sum_{\{s|cl.C \text{ matches } s\}} p(s)p(cl.A|s)}, \tag{4.9}$$

where $p(s)$ denotes the probability that state s is presented in the problem and $p(a|s)$ specifies the conditional probability that action (or classification) a is chosen given state s.

Given that reward prediction is well-estimated, we can derive the prediction error estimate in a similar fashion:

$$cl.\varepsilon \approx \frac{\sum_{\{s|cl.C \text{ matches } s\}} p(s)p(cl.A|s)(|cl.R - \rho(s, cl.A)|)}{\sum_{\{s|cl.C \text{ matches } s\}} p(s)p(cl.A|s)}. \tag{4.10}$$

Thus, the reward prediction estimates the mean reward encountered in a problem and the reward prediction error estimates the *mean absolute deviation* (MAD) from the reward prediction.

In a two-class classification problem, which provides 1000 reward if the chosen class was correct and 0 otherwise, we may denote the probability that a particular classifier predicts the correct outcome by p_c. Consequently, the reward prediction $cl.R$ of classifier cl will approximate $1000p_c$. Neglecting the

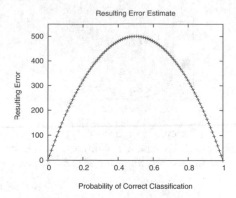

Fig. 4.2. Assuming a 1000/0 reward scheme, it can be seen that the prediction error estimate peaks at a probability of a correct classification of 0.5. Given that this is the probability of a completely general classifier, the more probable a classification is correct or wrong, the lower the error.

oscillation and consequently setting $cl.R$ equal to $1000p_c$, the following error derivation is possible:

$$cl.\varepsilon = (1000 - cl.R)p_c + cl.R(1 - p_c) =$$
$$= 2000(p_c - p_c^2). \tag{4.11}$$

The formula sums the two cases of executing a correct or wrong action with the respective probabilities. The result is a parabolic function for the error ε that reaches its maximum of 0.5 when $p_c(cl)$ equals 0.5 and is 0 for $p_c(cl) = 0$ and $p_c(cl) = 1$. The function is depicted in Figure 4.2. It should be noted that this parabolic function is very similar to the concept of entropy so that the prediction error estimate can also be regarded as an entropy estimate. The main idea behind this function is that given a 50/50 probability of classifying correctly, the mean absolute error will be on its highest value. The more consistently a classifier classifies problem instances correctly/incorrectly the lower the reward prediction error.

4.1.4 Rule Evolution

Besides the aforementioned covering mechanism that ensures that all actions in a particular problem instance are represented by at least one classifier, XCS applies a GA for rule evolution. Genetic reproduction is invoked in the current action set $[A]$ if the average time since the last GA application (stored in parameter ts) upon the classifiers in $[A]$ exceeds threshold θ_{GA}.

The GA selects two parental classifiers using proportionate selection where the probability of selecting classifier cl ($p_s(cl)$) is determined by its relative fitness in $[A]$, i.e. $p_s(cl) = F(cl)/\sum_{c \in [A]} F(c)$). Two offspring are generated reproducing the parents and applying crossover and mutation. Parents stay in the population competing with their offspring. Usually, we apply *free mutation*, in which each attribute of the offspring condition is mutated to the other two possibilities with equal probability. Another option is to apply *niche mutation*, which assures that the mutated classifier still matches the current problem instance.

Offspring parameters are initialized by setting prediction R to the currently received reward and reward prediction error ε to the average error in the action set. Fitness F and action set size estimate as are set to the parental values. Additionally, fitness F is decreased to 10% of the parental fitness being pessimistic about the offspring's fitness. Experience counter exp and numerosity num are set to one.

In the insertion process, *subsumption deletion* may be applied (Wilson, 1998) to stress generalization. Due to the possible strong effects of *action set subsumption* (Wilson, 1998; Butz & Wilson, 2002), we apply *GA subsumption* only. *GA subsumption* checks offspring classifiers to see whether their conditions are logically subsumed by the condition of another *accurate* ($\varepsilon < \varepsilon_0$) and *sufficiently experienced* ($exp > \theta_{sub}$) classifier in [A]. If an offspring is subsumed, it is not inserted in the population, but the subsumer's numerosity is increased.

The population of classifiers [P] is of maximum size N. Excess classifiers are deleted from [P] with probability proportional to the action set size estimate as that the classifiers occur in. If the classifier is sufficiently experienced $exp > \theta_{del}$ and its fitness F is significantly lower than the average fitness $\overline{F}$ of classifiers in [P] ($F < \delta\overline{F}$), its deletion probability proportion is increased by factor $\overline{F}/F$.

4.1.5 XCS Learning Intuition

The overall learning process is illustrated in Figure 4.3. As can be seen, in contrast to the simple LCS, XCS reproduces classifiers selecting from the current action set instead of from the whole population. Several additional classifier parameters, and most explicitly the accuracy-based fitness measure F, monitor the performance of the classifier.

XCS strives to predict all possible reward values equally accurately. The amount of reward itself does not bias XCS's learning mechanism. Additionally, XCS tends to evolve a *general* problem solution since reproduction favors classifiers that are frequently active (part of an action set) whereas deletion deletes from the whole population. Among classifiers that are *accurate* (that is, the error is below ε_0) and *semantically* equally general (the classifiers are active equally often), subsumption deletion causes additional generalization pressure in that a *syntactically* more general classifier absorbs more specialized offspring classifiers.

Due to the niche reproduction in conjunction with action set size based deletion, the evolving representation is designed to stay complete. Since the

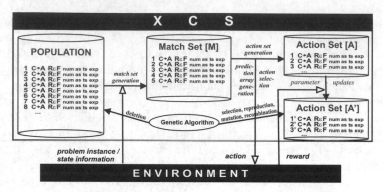

Fig. 4.3. Learning in XCS differs from simple LCSs in that the GA selects classifiers for reproduction in the current action set but for deletion in the whole population.

action set size estimate decreases in classifiers that currently inhabit underrepresented niches, their probability of deletion decreases, resulting in effective niching. Reproduction is dependent on the frequency of occurrence. In effect, XCS is designed to evolve accurate payoff predictions for all problem spaces that occur sufficiently frequently.

Together, the learning processes in XCS are designed to achieve one common goal: to evolve a *complete, maximally accurate, and maximally general* representation of the underlying payoff-map, or Q-value function. This representation was previously termed the optimal solution representation [O] (Kovacs, 1996; Kovacs, 1997).

Before we analyze the XCS system in detail in the subsequent chapters, the next section provides further insights into XCS, considering learning and problem representation of XCS in a small classification problem and a small RL problem.

4.2 Simple XCS Applications

In order to better understand XCS's mechanisms this section applies XCS to several small exemplar problems. In addition to the understanding of XCS functioning, this section provides a comparison to RL and the Q-learning approximation method in XCS. First, however, we investigate XCS's performance in a simple classification problem.

4.2.1 Simple Classification Problem

In order to disregard the additional complication of reward back-propagation, characterized by the Q term derived in Equation 4.2, a big part of the analyses in the subsequent chapters focuses on classification or one-step problems. Since in such problems successive problem instances usually depend neither on each other nor on the chosen classification, each learning iteration can be treated independently.

As our simple example, we use the 6-multiplexer problem (see Appendix C). The multiplexer problem is interesting because the solution can be represented by completely non-overlapping niches and the path to the solution is of interest as well. The optimal XCS population in the 6-multiplexer is shown in Table 4.1.

Table 4.1. The optimal population [O] in the 6-multiplexer problem.

Nr.	C	A	R	ε	F	Nr.	C	A	R	ε	F
1	000###	0	1000	0	1	2	000###	1	0	0	1
3	001###	0	0	0	1	4	001###	1	1000	0	1
5	01#0##	0	1000	0	1	6	01#0##	1	0	0	1
7	01#1##	0	0	0	1	8	01#1##	1	1000	0	1
9	10##0#	0	1000	0	1	10	10##0#	1	0	0	1
11	10##1#	0	0	0	1	12	10##1#	1	1000	0	1
13	11###0	0	1000	0	1	14	11###0	1	0	0	1
15	11###1	0	0	0	1	16	11###1	1	1000	0	1

It is important to notice that XCS represents both the correct classification and the incorrect classification for each problem subsolution. As mentioned before, XCS is designed to evolve a complete and accurate payoff map of the underlying problem. Thus, XCS represents each subsolution in the 6-multiplexer twice: (1) by specifying the correct condition and action combination; (2) by specifying the incorrect class as an action and predicting accurately ($\varepsilon = 0$) that the specified action is incorrect (resulting always in zero reward $R = 0$).

How might XCS evolve such an optimal population? In general, two ways may lead to success, starting either from the over-specific or the overgeneral side. Starting over-specific, the don't care probability $P_\#$ is set low so that initial classifiers nearly specify all l attributes in the problem. Most classifiers then are maximally accurate so that inaccurate classifiers quickly disappear since selection favors accurate classifiers. Mutation mainly results in generalized offspring since mainly specified attributes will be chosen for mutation. If mutation results in an inaccurate, overgeneralized classifier, the accuracy and thus fitness of the offspring is expected to drop and the classifier will have hardly any reproductive events and consequently will soon be deleted. Since more general classifiers will be more often part of an action set, more general, accurate classifiers undergo more reproductive events and thus propagate faster. Thus, the process is expected to evolve the accurate, maximally general solution as the final outcome.

Despite this appealing description, we will see that a start from the overspecialized side is usually undesirable because of the large requirements on population size. In essence, when starting completely specialized, the popula-

Table 4.2. The path to an optimal solution from the overgeneral side in the 6-multiplexer problem.

Nr.	C	A	R	ε	Nr.	C	A	R	ε
1	######	0	500.0	500.0	2	######	1	500.0	500.0
3	1#####	0	500.0	500.0	4	1#####	1	500.0	500.0
5	0#####	0	500.0	500.0	6	0#####	1	500.0	500.0
7	##1###	0	375.0	468.8	8	##1###	1	625.0	468.8
9	##0###	0	625.0	468.8	10	##0###	1	375.0	468.8
11	##11##	0	250.0	375.0	12	##11##	1	750.0	375.0
13	##00##	0	250.0	375.0	14	##00##	1	750.0	375.0
15	0#1###	0	250.0	375.0	16	0#1###	1	750.0	375.0
17	0#0###	0	750.0	375.0	18	0#0###	1	250.0	375.0
19	0#11##	0	0.0	0.0	20	0#11##	1	1000.0	0.0
21	001###	0	0.0	0.0	22	001###	1	1000.0	0.0
23	10##1#	0	0.0	0.0	24	10##1#	1	1000.0	0.0
25	000###	0	1000.0	0.0	26	000###	1	0.0	0.0
27	01#0##	0	1000.0	0.0	28	01#0##	1	0.0	0.0

tion size needs to be chosen larger than 2^l and thus exponentially in problem length, which is obviously a highly undesirable requirement.

Thus, XCS usually starts its search for an optimal solution representation from the overgeneral side. In the most extreme case, the don't care probability may be set to $P_\# = 1$ and XCS starts its search with the two completely general classifiers ######→0 and ######→1. Mutation is then initially responsible for the introduction of more specialized classifiers. However, mutation alone is usually not sufficient since more general classifiers are part of more action sets undergoing more reproductive events, on average resulting in an overall generalization pressure. Thus, classifiers need to be propagated that are more accurate. Table 4.2 shows the resulting expected average R and ε values for progressively more specialized classifiers. Fitness is not shown since fitness depends on the classifier distribution in the current population.

Note how initially the specialization of the value bits results in a smaller error and thus a larger accuracy. Once a value bit is specified, the specialization of an address bit or an additional value bit has an equal effect on error. Thus, the evolutionary process is initially guided towards specializing value bits. Soon, however, the beneficial specification of an address bit takes over and completely accurate classifiers evolve.

Note how XCS explores both sides of the solution spectrum: the more incorrect as well as the more correct classification side. In effect, a complete payoff map is evolved dependent on if better classifiers (lower error classifiers) have enough time to propagate and a complete solution can be supported. Subsequent chapters investigate these issues in further detail.

4.2.2 Performance in Maze 1

Compared to a classification problem, an RL problem poses the additional complication of back-propagating reward appropriately. Due to the generalized, distributed representation of the Q-function in XCS, additional complications might arise caused by inappropriate reward propagations from inaccurate, young, or overgeneralized classifiers. Chapter 10 investigates performance in RL problems in much more detail.

In this section, we investigate XCS learning and solution representation in the Maze 1 problem shown in Figure 2.3 on page 17. The question is, if XCS can be expected to solve this problem and evolve a complete, but generalized representation of the underlying Q-table.

Lanzi (2002) provides a detailed comparison of XCS with Q-learning from the RL perspective. The investigations show that if no generalization is allowed, XCS essentially mimics Q-learning. Each classifier corresponds to exactly one entry in the Q-table (shown in Table 2.2 on page 20 for the Maze 1 case) and the Q-learning update function:

$$Q(s, a) \leftarrow Q(s, a) + \beta(r + \gamma \text{max}_{a'} Q(s', a') - Q(s, a)) \tag{4.12}$$

is equivalent to the XCS update function for the reward prediction

$$R \leftarrow R + \beta(r + \gamma \text{max}_{A \in \mathcal{A}} P^{t+1}(A) - R) \tag{4.13}$$

in that the reward prediction values R coincide with the Q-values $Q(s, a)$ since the condition of a classifier specifies exactly one state s and one action a. Thus, the prediction array coincides with Q-value entries, since for each prediction array entry exactly one classifier applies, so that the entry is equivalent to the R value of the classifier (see Equation 4.1), which is equivalent to the Q-value as outlined above. Thus, without generalization, XCS is a tabular Q-learner where each tabular entry is represented by a distinct rule.

What is expected to happen in the case when generalization is allowed? What does the perfect representation look like? The optimal population in the Maze 1 problem is shown in Table 4.3. In comparison to the Q-table in Table 2.2, we see that XCS is able to represent the maze in a much more compact form requiring only 19 classifiers to represent a Q-table of size 40. Note how XCS is generalizing over the state space with respect to each action as indicated by the action order in Table 4.3. Essentially, since each state combination is representable with a different action code (for example, the condition of classifier one specifies that it matches in situations A or E), XCS generalizes over states that yield identical Q-values with respect to a specific action. Even in our relatively simple environment in which there are only three different Q-values possible (810, 900, and 1000) generalization is effective. Given a larger state space, further generalizations over different states are expectable. In sum, XCS learns a generalized Q-table representation specializing those attributes that are relevant for the distinction of different Q-values.

Table 4.3. The optimal population $[O]$ in the Maze 1 problem.

Nr.	C	A	R	ε	Nr.	C	A	R	ε
1	11#####1	↑	810	0	2	1#0###0#	↑	900	0
3	0#######	↑	1000	0					
4	11#####1	↗	810	0	5	#10###0#	↗	900	0
6	#0######	↗	1000	0					
7	##0####1	→	900	0	8	11####0#	→	810	0
9	11#####1	↘	810	0	10	##0###0#	↘	900	0
11	11#####1	↓	810	0	12	##0###0#	↓	900	0
13	11#####1	↙	810	0	14	##0###0#	↙	900	0
15	1#0####1	←	810	0	16	#1####0#	←	900	0
17	11#####1	↖	810	0	18	##0###01	↖	900	0
19	#######0	↖	1000	0					

The question remains how XCS learns the desired complete, accurate, and maximally problem solution shown in Table 4.3. Dependent on the initialization procedure, classifiers might initially randomly distinguish some states from others. Additionally, the back-propagated reward signal is expected to fluctuate significantly. As in Q-learning, XCS is expected to progressively learn starting from those state-action combinations that yield actual reinforcement. Since these transitions result in an exact reward of 1000, classifiers that identify these cases quickly become accurate and thus will be reproduced most of the time they are activated. Once, the 1000 reward cases are stably represented, back-propagation becomes more reliable and the next reward level (900) can be learned accurately and so forth. Thus, as in Q-learning, reward will be spread backward starting from the cases that yield actual reward. In chapter 10, we show how XCS's performance can be further improved to ensure a more reliable and stable solution in RL problems applying gradient-based update techniques. In sum, due to the accuracy-based fitness approach, XCS is expected to evolve a representation that covers the whole problem space of Maze 1, yielding classifiers that comprise the maximal number of states in their conditions to predict the Q-value accurately with respect to the specified action.

We still have not answered how XCS evolves a *maximally general* problem representation. For example, Classifier 12 specifies that if moving south and if currently located in state B, C, or D, then a reward of 900 is expected. Classifier 12 is *maximally general* in that any further generalization of the classifier's condition will result in an inaccurate prediction (that is, state A or E will be accepted by the condition part in which a south action yields a reward of 810). However, Classifier 12 can be further specialized still matching in all three states B, C, and D. In essence, in all three states the positions to the southeast, south, and southwest are blocked by obstacles so that the specification of obstacles in these positions is redundant. Only subsumption is able to favor the syntactically more general Classifier 12. Thus, due to the sparse problem instance coverage (only five perceptions possible) of the whole problem space (the coding allows $2^l = 2^8 = 256$ different problem instances), only few (semantic) generalizations over different states are possible. Syntactic generalizations, which identify redundant specializations, are much more important. Thus, subsumption is a major factor for evolving maximally general classifiers in environments with sparse problem input space S coverage. If Maze 1 was noisy and syntactic generalization was desired, the error threshold ε_0 may need to be increased to enable subsumption, as suggested in Lanzi (1999c).

4.3 Summary and Conclusions

This chapter introduced the accuracy-based learning classifier system XCS. As all (Michigan-style) LCSs, XCS is an online learner applying RL techniques for rule evaluation and action decisions and GA techniques for rule discovery. In contrast to traditional LCSs, XCS derives classifier fitness from the *accuracy of reward prediction* instead of from the reward prediction value directly. The reward updates can be compared to Q-learning updates. Thus, XCS is designed to learn a generalized representation of the underlying Q-function in the problem.

XCS's *niche reproduction* in combination with population-wide deletion results in an *implicit, semantic generalization pressure* that favors classifiers that are more frequently active. Additionally, subsumption deletion favors accurate classifiers that are *syntactically* more general. Niche reproduction in combination with population-wide deletion based on the action set size estimates results in effective problem niching striving to evolve and maintain a complete problem solution.

In effect, XCS is designed to evolve a *complete, maximally accurate, and maximally general problem solution* represented by a population of classifiers. Each classifier defines a problem subspace in which it predicts the corresponding Q-value accurately.

The application to two small problems showed which solution representation XCS is designed to evolve and how it might be evolved. Starting from the overgeneral side, higher accurate classifiers need to be detected and propagated. Starting from the over-specific side, generalizations due to mutations boil the representation down to the desired accurate, maximally general one. However, when starting over-specialized, the required population size needs to grow exponentially in problem length l given that all problem instances are equally likely to occur.

In problems in which only a few samples are available so that the problem space is covered only sparsely, syntactic generalization and thus subsumption becomes more important since the usual generalization pressure due to more frequent reproductive opportunities may not apply. However, subsumption only works for accurate classifiers so that a proper choice of the error threshold ε_0, and thus an appropriate noise tolerance, becomes more important.

While the base XCS system introduced in this section learns a generalized representation of the underlying Q-value function, it should be noted that XCS is not limited to approximating Q-functions. In fact, XCS can be modified yielding a general, online learning *function approximation technique* (Wilson, 2001a). Due to the rule-based structure, XCS is designed to partition its space dependent on the representation of classifier conditions. Each classifier then approximates, or predicts, the actual function value in its defined subspace. In the base XCS, conditions define hyperrectangles as subspaces and predict constant reward values—consequently applying piece-wise constant function approximation. Wilson (2001a) experimented with piecewise linear approximations. Other prediction methods are imaginable as well, such as the most recently introduced polynomial predictions (Lanzi, Loiacono, Wilson, & Goldberg, 2005). Similarly, XCS can be endowed with other condition structures such as S-expressions (Lanzi, 1999b). As we will see, dependent on the problem structure at hand, the most suitable condition structure can be expected to be most advantageous for learning with XCS.

5

How XCS Works: Ensuring Effective Evolutionary Pressures

The last chapter gave a concise introduction to the accuracy-based XCS classifier system. We saw that XCS is designed to evolve online a complete, maximally accurate, and maximally general solution to the problem at hand (e.g. by approximating the Q-value function). The accuracy-based approach assures that no strong overgenerals are possible since the maximally accurate classifiers receive maximal fitness.

With this knowledge in mind, we now turn to the further aspects of our facetwise LCS theory approach regarding the evolutionary pressures in the XCS classifier system. While strong overgenerals are prevented by the accuracy-based fitness approach, it still needs to be assured that fitness *guides* towards the intended solution. Second, parameter initialization and estimation needs to be most effective. Third, appropriate generalization needs to apply so that the solution becomes maximally general.

To investigate these points in XCS, we undertake a general analysis of all *evolutionary pressures* in XCS. Evolutionary pressures can be regarded as evolutionary biases that influence or bias learning in XCS. Often, the pressures influence *specificity* in the classifier population. Specificity was defined in Chapter 3. It essentially characterizes how restricted a classifier condition is. The smaller the problem subspace a classifier condition covers, the more specific it is.

The pressure analysis quantifies generalization in XCS as well as the influence of mutation. It leads to an equilibrium in population specificity if no fitness pressure applies. This interaction is quantified in the *specificity equation*, which we evaluate in detail. With the addition of fitness pressure and subsumption, the equilibrium lies exactly at the point of the desired maximally accurate and maximally general problem solution.

However, in order to ensure that fitness guides to the equilibrium, the generalization pressure needs to be overcome. This is assured by an appropriate selection mechanism leading us to the introduction of *tournament selection* for offspring selection. We show that tournament selection assures sufficiently strong fitness guidance and consequently makes XCS much more parameter as well as noise independent.

The next section introduces all evolutionary pressures, analyzes them in separation, and combines them in one general specificity equation. The results are experimentally validated in binary classification problems. Next, tournament selection is introduced and evaluated. The investigations prepare XCS to face the subsequent challenges outlined in the facetwise theory approach.

5.1 Evolutionary Pressures in XCS

Previous publications have considered the influence of fitness guidance and generalization. The principles underlying evolution in XCS were originally outlined in Wilson's *generalization hypothesis* (Wilson, 1995), which suggests that classifiers in XCS become maximally general due to niche-based reproduction in combination with population-wide deletion. Subsequently, Kovacs (1996) extended Wilson's explanation to an *optimality hypothesis*, supported experimentally in small multiplexer problems, in which he argued that XCS develops minimal representations of optimal solutions (that is, the optimal solution representation [O]). Later, Wilson (1998) suggested that XCS scales polynomially in problem complexity and thus is machine learning competitive. Kovacs and Kerber (2001) related problem complexity directly to the size of the optimal solution ||[O]||. However, Butz, Kovacs, Lanzi, and Wilson (2001) showed that the ||[O]|| measure is clearly not the only measure that influences XCS's learning behavior.

Despite these insights, Wilson's generalization hypothesis still needed to be theoretically validated and quantified. In this section, we investigate Wilson's hypothesis developing a fundamental theory of XCS generalization and learning. To avoid the additional complication of back-propagating reward in RL problems, we focus on XCS's performance in classification problems.

In particular, we analyze the evolutionary pressures present in XCS. An evolutionary pressure refers to a learning bias in XCS influencing the population structure, often with respect to specificity. The average *specificity* in a population, denoted by $\sigma[P]$, refers to the average specificity of all classifiers in the population. That is,

$$\sigma[P] = \frac{\sum_{c \in [P]} \sigma(c) \cdot c.num}{\sum_{c \in [P]} c.num}, \tag{5.1}$$

where $\sigma(c)$ refers to the specificity of classifier c. We defined specificity for the binary case in Chapter 3 as the fraction of specified attributes (zero or one) in the condition part of a classifier.

Our analysis distinguishes between the following evolutionary pressures:

1. *Set pressure*, which quantifies Wilson's generalization hypothesis;

2. *Mutation pressure*, which quantifies the influence of mutation on specificity;
3. *Deletion pressure*, which qualifies additional deletion influences;
4. *Fitness pressure*, which qualifies the accuracy-based fitness influence;
5. *Subsumption pressure*, which qualifies the exact influence of subsumption deletion.

Set pressure and mutation pressure are combined to a general *specificity equation* that is evaluated in several experiments progressively increasing the influence of deletion and fitness pressure.

5.1.1 Semantic Generalization due to Set Pressure

The basic idea behind the set pressure is that XCS reproduces classifiers in action sets $[A]$ whereas it deletes classifiers from the whole population $[P]$. The set pressure is a combination of the selection pressure produced by the GA applied in $[A]$ and the pressure produced by deletion applied in $[P]$. It was originally qualitatively proposed in Wilson's original paper (Wilson, 1995) and later further experimentally analyzed by Kovacs (1996).

The generalization hypothesis argues that since more general classifiers appear more often in action sets $[A]$, they undergo more reproductive events. Combined with deletion from $[P]$, the result is an intrinsic tendency towards generality favoring more general classifiers. Classifiers in this respect are *semantically* more general in that they are more frequently part of an action set. Classifiers that are equally often part of an action set but may be distinguished by syntactic generality are not affected by the set pressure.

To formalize the set pressure, we determine the expected specificity $\sigma[A]$ of classifiers in an action set $[A]$ with respect to the current expected specificity $\sigma[P]$ of the classifiers in population $[P]$. The specificity of the initial random population $\sigma[P]$ is directly correlated with the don't-care probability $P_\#$, i.e., $\sigma[P] = 1 - P_\#$. For our calculations, we assume a binomial specificity distribution in the population. This assumption essentially holds in the case of a randomly generated population. It enables us to determine the probability that a randomly chosen classifier in the population $cl \in [P]$ has specificity k/l as follows:

$$P(\sigma(cl) = k/l | \sigma[P]) = \binom{l}{k} \sigma[P]^k (1 - \sigma[P])^{l-k}, \qquad (5.2)$$

where cl is a classifier; l is the length of classifier conditions; and k is the number of *specified attributes* in the condition, that is, the number of attributes different from a don't-care symbol. The equation essentially is able to estimate the proportion of different specificities in a population with average specificity $\sigma[P]$.

The probability that a classifier cl matches a certain input s depends on its specificity $\sigma(cl)$. To match, a classifier cl with specificity k/l must match all k specific bits. This event has probability 0.5^k since each specific attribute matches with probability 0.5. Therefore, the proportion of classifiers in [P] with a specificity k/l that match in a specific situation is as follows:

$$P(cl\ matches \wedge \sigma(cl) = k/l | \sigma[P]) =$$
$$= P(\sigma(cl) = k/l | \sigma[P]) P(cl\ matches | \sigma(cl) = k/l) =$$
$$= P(\sigma(cl) = k/l | \sigma[P]) 0.5^k = \binom{l}{k} \left(\frac{\sigma[P]}{2} \right)^k (1 - \sigma[P])^{l-k} \qquad (5.3)$$

To derive a specificity $\sigma[M]$ of a match set $[M]$, it is first necessary to specify the proportion of classifiers in $[M]$ with specificity k/l given the population specificity $\sigma[P]$. This proportion can be derived as follows:

$$P(\sigma(cl) = k/l | cl \in [M] \wedge \sigma[P]) = \frac{P(cl\ matches \wedge \sigma(cl) = k/l) | \sigma[P]}{\sum_{i=0}^{l} P(cl\ matches \wedge \sigma(cl) = i/l) | \sigma[P]} =$$

$$= \frac{\binom{l}{k} \left(\frac{\sigma[P]}{2} \right)^k (1 - \sigma[P])^{l-k}}{\sum_{i=0}^{l} \binom{l}{i} \left(\frac{\sigma[P]}{2} \right)^i (1 - \sigma[P])^{l-i}} = \frac{\binom{l}{k} \left(\frac{\sigma[P]}{2} \right)^k (1 - \sigma[P])^{l-k}}{\left(1 - \frac{\sigma[P]}{2} \right)^l} =$$

$$= \binom{l}{k} \left(\frac{\sigma[P]}{2 - \sigma[P]} \right)^k \left(1 - \frac{\sigma[P]}{2 - \sigma[P]} \right)^{l-k} \qquad (5.4)$$

To compute $\sigma[M]$ we multiply actual specificity values, k/l, by the proportions $P(\sigma[M] = k/l | \sigma[P])$ and sum up the values to derive the resulting specificity of $[M]$. Since the action set $[A]$ has on average the same specificity as the match set $[M]$ ($\sigma[A] \approx \sigma[M]$), $\sigma[A]$ can be derived as follows:

$$\sigma[A] \approx \sigma[M] =$$

$$= \sum_{k=0}^{l} \frac{k}{l} P(\sigma(cl) = k/l | cl \in [M] \wedge \sigma[P]) =$$

$$= \sum_{k=1}^{l} \frac{k}{l} \binom{l}{k} \left(\frac{\sigma[P]}{2 - \sigma[P]} \right)^k \left(1 - \frac{\sigma[P]}{2 - \sigma[P]} \right)^{l-k} =$$

$$= \sum_{k=1}^{l} \binom{l-1}{k-1} \left(\frac{\sigma[P]}{2 - \sigma[P]} \right)^k \left(1 - \frac{\sigma[P]}{2 - \sigma[P]} \right)^{l-k} =$$

$$= \frac{\sigma[P]}{2 - \sigma[P]} \sum_{j=0}^{l-1} \binom{l-1}{j} \left(\frac{\sigma[P]}{2 - \sigma[P]} \right)^j \left(1 - \frac{\sigma[P]}{2 - \sigma[P]} \right)^{l-1-j} =$$

$$= \frac{\sigma[P]}{2 - \sigma[P]} \qquad (5.5)$$

The equation can be used to determine the average expected specificity $\sigma[A]$ in an action set $[A]$ assuming a binomially distributed specificity with mean $\sigma[P]$ in the population.

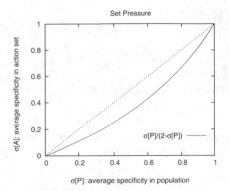

Fig. 5.1. Except at the lower and upper bounds, the expected average specificity of action sets $\sigma[A]$ is always smaller than that of the current population.

Figure 5.1 depicts Equation 5.5. Except at the lower and upper bounds, the specificity of $[A]$ is always smaller than the specificity of $[P]$. Thus, since selection takes place in the action sets but deletion occurs in the population as a whole, there is a tendency for the generality of the population to increase—in line with Wilson's generalization hypothesis. In the absence of fitness pressure, the equation provides an estimate of the difference in specificity of selected and deleted classifiers. Equation 5.5 is enhanced below accounting for mutation as well.

5.1.2 Mutation's Influence

Although usually only a low mutation probability is applied, mutation still influences specificity. In the absence of other evolutionary influences, mutation pushes the population towards a certain proportion of zeros, ones, and don't-cares. As outlined in Chapter 3, free mutation pushes towards a distribution of 1:2 general:specific. *Niche mutation*, which mutates a specified attribute always to a don't care and a don't care always to the current value of the respective attribute, pushes towards a distribution of 1:1 general:specific.

The average expected change in specificity between the parental classifier c_p and the mutated offspring classifier c_o for the niche mutation case can be written as follows:

$$\Delta_{mn}(\sigma(c_p)) = \sigma(c_o) - \sigma(c_p) =$$
$$= \sigma(c_p)(1 - \mu) + (1 - \sigma(c_p))\mu - \sigma(c_p) =$$
$$= \mu(1 - 2\sigma(c_p)), \tag{5.6}$$

and for free mutation as

$$\Delta_{mf}(\sigma(c_p)) = \sigma(c_o) - \sigma(c_p) =$$
$$= \sigma(c_p)(1 - \frac{\mu}{2}) + (1 - \sigma(c_p))\mu - \sigma(c_p) =$$
$$= 0.5\mu(2 - 3\sigma(c_p)). \tag{5.7}$$

As expected, the increase in specificity is higher when free mutation is applied given low parental specificity ($\sigma(c_p) < 1/2$) and specificity decrease is lower when parental specificity is high ($\sigma(c_p) > 2/3$). Applying random selection, mutation, and random deletion, mutation pushes the population towards a specificity of 0.5 applying niche mutation and 0.66 applying free mutation. The current intensity of the pressure depends on the mutation type, the current parental specificity, and on the frequency of the GA application (influenced by the parameter θ_{GA}).

5.1.3 Deletion Pressure

The probability of a classifier being deleted depends on its action set size estimate as and (depending on classifier experience) its fitness F. Due to the resulting bias towards deleting classifiers that occupy larger action sets, deletion pushes the population towards an equal distribution of classifiers in each environmental niche. With respect to specificity, classifier selection for deletion from $[P]$ is essentially random and there is no particular deletion pressure for or against general classifiers. In the absence of other biases the average expected specificity of deleted classifiers is equal to the average specificity in the population $\sigma[P]$.

A more significant effect can be observed with respect to overlapping niches. Given there are two non-overlapping, accurate niches and another accurate niche that overlaps with either one of the former, the action set size estimate of the overlapping niche will be larger than that of the non-overlapping ones. For example, given the non-overlapping niches 000*** and 01*0** and the overlapping niche 0*00** (this is actually the case in the multiplexer problem), and given further that all niches are represented by accurate, maximally general classifiers with a numerosity of say ten, then the action set size estimate of the overlapping classifier will stay on 20 whereas the estimate of the non-overlapping ones will approximate 15, making the deletion of the overlapping classifier more likely. Thus, apart from emphasizing equal niche support, the action-set size estimate based deletion pushes the population towards a non-overlapping solution representation.

5.1.4 Subsumption Pressure

Subsumption deletion applies only to classifiers that are accurate ($\varepsilon < \varepsilon_0$) and sufficiently experienced ($exp > \theta_{sub}$). The accuracy requirement suggests that the problem is less noisy than ε_0 since otherwise classifiers are not expected to satisfy the criterion ever.

If accurate classifiers evolve, subsumption deletion pushes towards maximal *syntactic* generality in contrast to the set pressure which pushes towards *semantic* generality. *GA subsumption deletion* prevents the insertion of offspring into [P], if there is a classifier in [A] that is more general than the generated offspring. Thus, once an accurate, maximally general solution was found for a particular niche, no accurate, more specialized classifier will be inserted anymore, disabling any specialization in the current niche.

To summarize, subsumption pressure is an additional pressure towards accurate, maximally syntactically general classifiers from the over-specific side. It applies only when accurate classifiers are found. Thus, subsumption pressure is helpful mainly later in the learning process once accurate classifiers are found. It usually results in a strong decrease of population size, focusing on maximally general classifiers.

5.1.5 Fitness Pressure

Until now we have not considered the effect of fitness pressure which can influence several other pressures. Fitness pressure is highly dependent on the particular problem being studied and is therefore difficult to formalize. In general, fitness results in a pressure which pushes [P] from the overgeneral side towards accurate classifiers. Late in the run, when the optimal solution is mainly found, it prevents overgeneralization.

As in the case of subsumption, fitness also pushes the population towards a non-overlapping problem representation. Since fitness is derived from the relative accuracy, the accuracy-share is lower in classifiers that overlap with many other accurate classifiers. In effect, unnecessary, overlapping classifiers have a lower fitness on average, are thus less likely to reproduce, and are thus likely to be deleted from the population.

In terms of specificity, fitness pressure towards higher accuracy results usually in a specialization pressure since higher specificity usually implies higher accuracy. Certain problems, however, may mislead fitness guidance in that more general classifiers may actually have higher accuracy. This is particularly the case in problems with unbalanced class distributions and multiple classes (Butz, Goldberg, & Tharakunnel, 2003; Bernadó-Mansilla & Garrell-Guiu, 2003). Since the problem is not as severe as originally suspected, we do not investigate it any further. As suggested by our facetwise theory, though, fitness guidance needs to be ensured.

In sum, fitness pressure usually works somewhat in the opposite direction (towards higher specificity) of the set pressure. Thus, given fitness pressure in a problem, the specificity in the population is expected to decrease less or to increase dependent on the amount of fitness pressure. Fitness pressure is certainly highly dependent on the investigated problem and thus hard to quantify. The following section combines the pressure influences into the specificity equation.

5.1.6 Pressure Interaction

We now combine the above evolutionary pressures and analyze their inter-action. Initially, we consider the interaction of set pressure, mutation pressure, and deletion pressure, which yields an important relationship we call the *specificity equation*. Next, we consider the effect of subsumption pressure and potential fitness influences. Finally, we provide a visualization of the interaction of all the pressures. The analyses are experimentally evaluated in Section 5.1.7.

Specificity Equation

Since mutation is only dependent on the specificity of the selected parental classifier and deletion can be assumed to be random, selection and mutation can be combined into one *specificity equation*. Essentially, set pressure generalizes whereas mutation specializes *or* generalizes dependent on the specificity of the currently selected classifier.

Since fitness pressure is highly problem dependent, we disregard fitness influences, essentially assuming equal fitness of all classifiers in our analysis. As shown later in Section 5.1.7, this assumption *holds* when all classifiers are accurate and nearly holds when all are similarly inaccurate. Despite the fitness equality assumption, deletion is also dependent on the action set size estimate *as* of a classifier. However, in accordance with Kovacs' insight on the relatively small influence of this dependence (Kovacs, 1999), we assume a random deletion from the population in our formulation. Thus, as stated above, a deletion results on average in the deletion of a classifier with a specificity equal to the specificity of the population $\sigma[P]$. The generation of an offspring, on the other hand, results in the insertion of a classifier with an average specificity of $\sigma[A] + \Delta_{mx}(\sigma[A])$ $(x \in f, n)$ dependent on the type of mutation used. Putting the observations together we can calculate the average specificity of the resulting population after one time step:

$$\sigma[P(t+1)] = \sigma[P(t)] + f_{GA} \frac{2(\sigma[A] + \Delta_{mx}(\sigma[A]) - \sigma[P(t)])}{N} \qquad (5.8)$$

The parameter f_{GA} denotes the frequency of a GA application per time step assuming a constant application frequency for now. The formula adds to the current specificity in the population $\sigma[P(t)]$ the expected change in specificity calculated as the difference between the specificity of the two reproduced and mutated classifiers, that is, $\sigma[A] + \Delta_{mx}(\sigma[A])$ and $\sigma[P(t)]$. Note that although the frequency f_{GA} is written as a constant in the equation, f_{GA} actually depends on θ_{GA}, the specificity $\sigma[P(t)]$, as well as on the specificity distribution in the population. Thus, f_{GA} cannot be written as a constant in general. However, by setting θ_{GA} to zero, it is possible to force f_{GA} to be one since the average time since the last application of the GA in an action set will always be at least one.

XCS's tendency towards accurate, maximally general classifiers from the overgeneral side is not dependent on the use of subsumption. However, subsumption is helpful in focusing the population more on the maximally general representation. In fact, although the set pressure pushes the population towards more general classifiers the pressure is somewhat limited. Equation 5.8 shows that without subsumption the complete convergence of the population towards maximally accurate, maximally general classifiers is not assured. However, XCS is an online learning system that should be flexible with respect to problem dynamics so that complete convergence is usually not desired.

Another reason for a potential lack of complete convergence can be that the set pressure is not present at all. This can happen if the state space of a problem is a proper subspace of all possible representable states $\{0,1\}^l$ (as is essentially the case in most datamining applications as well as in RL problems). Subsumption can be helpful in generalizing the population further.

As mentioned above, fitness pushes towards higher specificity from the overgeneral side. Equation 5.8 assumes that the selected parental classifier has an expected average specificity of $\sigma[A]$, effectively assuming random selection in the action set. However, selection is biased towards the selection of more accurate classifiers. In effect, $\sigma[A]$ needs to be replaced by the expected offspring specificity that depends on the expected fitness distribution in the action set. Since this distribution is not only dependent on the problem but also on the selection method used and the current specificity distribution in each action set, we won't analyze the fitness influence any further. However, it should be kept in mind that fitness influences are expected to cause a specialization pressure that diminishes the generalization effect of the set pressure.

All Pressures

The interaction of all the pressures is illustrated in Figure 5.2. In particular, the fitness pressure pushes $[P]$ towards more accurate classifiers; the set pressure pushes $[P]$ towards more general classifiers; the subsumption pressure pushes $[P]$ towards classifiers that are accurate and syntactically maximally general; the mutation pressure pushes towards a fixed proportion of symbols in classifier conditions. Deletion pressure is implicitly included in the notion of set pressure. More detailed effects of deletion are not depicted. Overall, these pressures lead the population towards a population of accurate maximally general classifiers. While set pressure and mutation pressure (free mutation is represented) are independent of classifier accuracy, subsumption pressure and, of course, fitness pressure are influenced by accuracy.

5.1.7 Validation of the Specificity Equation

This section evaluates the specificity equation, formulated in Equation 5.8, and the additional evolutionary pressures identified in the previous sections.

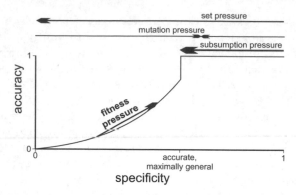

Fig. 5.2. The sketched interaction of all evolutionary pressures on imaginary specificity and accuracy axes shows how the evolutionary process in XCS is designed to evolve accurate, maximally general classifiers. If the classifiers are over-general (that is, specificity is smaller than at the accurate, maximally general point), accuracy is low (strongly dependent on the problem and XCS's accuracy equation) and fitness pushes towards the propagation of more accurate classifiers. On the other hand, if classifiers are over-specialized, accuracy is maximal but further generalization is possible. Set pressure and subsumption pressure are responsible to stress such generalizations.

The specificity equation summarizes the effect of three main evolutionary pressures in XCS: set pressure , mutation pressure, and deletion pressure.

XCS is applied to Boolean strings of length $l = 20$ with different settings. The following figures show runs with mutation rates varying from 0.02 to 0.20. In each plot, solid lines denote the result from Equation 5.8; while crossed lines represent the result of actual XCS runs. Curves are averages over 50 runs. If not stated differently, the population is initially filled up with random classifiers with don't-care probability $P_\# = 0.5$. Niche mutation is applied. The other XCS parameters are set as follows: $N = 2000$; $\beta = 0.2$; $\alpha = 0.1$; $\varepsilon_0 = 10$; $\nu = 5$; $\theta_{GA} = 0$; $\chi = 0.8$, $\theta_{del} = 20$; $\delta = 0.1$; and $\theta_{sub} = \infty$. Note that the discount factor γ is irrelevant here since these are classification or single-step problems. Since this section is concerned with the set pressure, subsumption is turned off to prevent additional generalization influences.

Constant Function

We begin the validation of Equation 5.8 examining runs in a constant function, which always returns one. With these settings, all classifiers turn out to be accurate since their prediction error is always zero. Note however that a zero prediction error does not necessarily mean constant fitness values. In fact, since fitness is determined by the classifier's *relative accuracy*, fitness should still have an influence on evolutionary pressure. We investigate the influence of free mutation, niche mutation, the GA threshold θ_{GA}, and the influence of the deletion pressure.

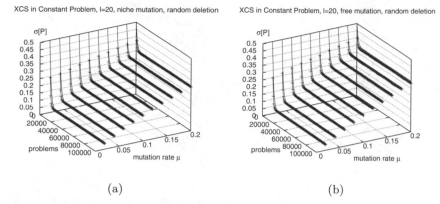

Fig. 5.3. Solid lines represent the specificity as predicted in Equation 5.8. Marked lines represent the actual experimental specificity $\sigma[P]$. In a constant function setting, the actual specificity behaves nearly exactly as predicted by the model applying either (a) niche mutation or (b) free mutation.

Niche Mutation.

Figure 5.3a) depicts the average specificity $\sigma[P]$ in the specified settings varying the mutation rate. The deletion pressure is eliminated by deleting classifiers uniformly randomly in the population based solely on the numerosity *num*.

   The empirical specificities match very closely to the model expressed in Equation 5.8. The initial specificity of 0.5 drops off quickly in the beginning due to the strong set pressure. Soon the specializing effect of the mutation pressure becomes visible and the specificity in the population converges as predicted. The higher the mutation rate μ, the stronger the specializing influence of mutation, as manifested in the higher convergence value in the curves with higher μ.

Free Mutation.

Figure 5.3b) depicts the specificity of the population [P] when free mutation is used. Besides the visibility of the mutation pressure due to the variation of μ, Figure 5.3b) confirms that free mutation has a slightly stronger influence on specificity as formulated in Equation 5.7. When directly comparing Figures 5.3a) and 5.3b) we note that the higher the parameter μ, the higher the influence of mutation pressure and thus the higher the differences in specificity due to the different mutation types.

θ_{GA} Threshold.

As noted above, the GA frequency f_{GA} in Equation 5.8 can generally not be written as a constant value. The frequency depends on the specificity $\sigma[P(t)]$

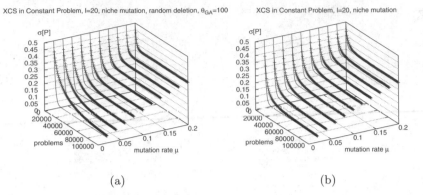

(a)   (b)

Fig. 5.4. (a) When applying a GA threshold of $\theta_{GA} = 100$, the GA frequency and consequently the specificity pressure decreases. Convergence values are not influenced. (b) Due to slow adaptation of the action set size estimate parameter, specificity convergence takes longer when action set size estimate based deletion is applied.

of the population, its distribution, and the problem distribution. However, the GA frequency f_{GA} equals one when the GA threshold θ_{GA} is set to zero (as has been done in the above evaluations). When setting θ_{GA} to a higher value (e.g. $\theta_{GA} = 100$), Figure 5.4a) reveals the lower GA frequency effect. Once the specificity in the population has dropped, the action set sizes increase since more classifiers match a specific state. Consequently, more classifiers take part in a GA application, more time stamps ts are updated, the average time since the last GA application in the population and in the action sets decrease, and finally the GA frequency decreases. The decrease is observable in the slower specificity decrease. As predicted by Equation 5.8, f_{GA} does not influence the convergence value.

Normal Deletion

The behavior of $\sigma[P]$ changes when we apply the usual deletion mechanism, which deletes proportional to the action set size estimate parameter as. Figure 5.4b) reports runs in which the usual deletion is used. Note that also in Figure 5.4b) the slopes of the curves decrease in comparison to the ones in Figure 5.3a). In the end, though, specificity of $[P]$ converges to the value predicted by the theory. The difference can only be the result of the bias in the deletion method of deleting classifiers with larger action set size estimates as. As the specificity of $[P]$ decreases, the action set size increases as noted before. Thus, since more general classifiers are more often present in action sets, their action set size estimate as is more sensitive to the change in the action set size and consequently, it is larger in more general classifiers while specificity drops. Eventually, all as values will have adjusted to the change

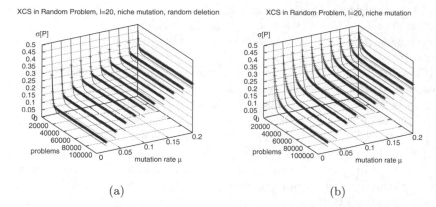

Fig. 5.5. (a) Applied to a random function, which returns zero or one equally probable for each problem instance, the specificity stays on a higher level due to the higher error variance in more specialized classifiers and parameter initialization effects. (b) Fitness-biased deletion decreases convergence of specificity due to the discussed action set size estimate influence.

and the predicted convergence value is met. This explanation is further confirmed by the fact that the difference between the actual runs and the curves given by Equation 5.8 become smaller and converge faster for higher mutation rates μ since the specificity slope is not as steep as in the curves with lower μ values.

Random Function

The results in the previous section show that the specificity equations accurately predict the behavior of XCS's average specificity. The fitness influence in XCS with a constant function proved to be rather small. Accordingly, we now apply XCS with two different deletion strategies to a much more challenging problem, that is, to a random Boolean function, which randomly returns rewards of thousand and zero. Figure 5.5a) reports the runs in which XCS with random deletion is applied to the random function. The experiments show that in the case of a random function, the fitness influences the specificity slope as well as the convergence value. In fact, the convergence value is larger than predicted by the model in Equation 5.8. The main effect is caused by the chosen initialization technique as well as by the higher error, and thus fitness variance in less experienced classifiers.

Since the possible rewards are 0 and 1000, assuming accurate parameter estimates in a classifier, classifier predictions fluctuate around 500, and consequently also the prediction errors fluctuate around 500. As in the more sensitive action set size estimates in Section 5.1.7, here the sensitivity is manifested in the prediction error ε. More specific classifiers have a less sensitive

ε and consequently a higher variance in the ε values. Since the accuracy calculation expressed in Equation 4.5 scales the prediction error to the power ν, the higher variance causes an *on average* higher accuracy and thus higher fitness.

Different parameter initialization techniques in combination with the moyenne adaptative modifiée technique enhance this influence. The more a classifier is inexperienced, the more the classifier parameters are dependent on the most recent cases. This, in combination with the scaled fitness approach, can make the effect even stronger. Since we set the experience exp of a new classifier to 1, XCS keeps the decreased parental parameter estimates so that fitness over-estimation is prevented. In fact, experimental runs with $exp = 0$ show that the specificity can increase to a level of even 0.2 independent of the mutation rate and type.

When applying the usual deletion strategy, based on as and the fitness estimate F, deletion causes an increase in the specificity of $[P]$ early on as shown in Figure 5.5b). This longer convergence time is attributable to the bias on as as already observed in Figure 5.4b). The additional fitness bias causes hardly any observable influence.

Overall, it can be seen that in a random function fitness causes an intrinsic pressure towards higher specificity. This pressure is due to the parameter initialization method and the higher variance in more specific classifiers. The on-average higher fitness in more specific classifiers causes fitness pressure and deletion pressure to favor those more specific classifiers; thus the resulting undirected slight pressure towards higher specificity. Note that the specificity change and the convergence to a particular specificity level observed in Figure 5.5b) should essentially take place in all problems that are similar to a random function. This is particularly the case if classifiers are overgeneral and the investigated problem provides no fitness guidance from the overgeneral side.

Steady State Specificity Distribution

The performance evaluations showed that the specificity equation is correct. Initialization influences and variance effects usually result in a slightly higher specificity than predicted.

From Equation 5.8 it is possible to derive the specificity a population is expected to converge to (regardless of the initial specificity) assuming no fitness influence. Setting the difference $(\sigma[A]) + \Delta_{mx}(\sigma[A]) - (\sigma[P])$ to zero, we derive for free mutation

$$\sigma[A] + \Delta_{mf}(\sigma[A]) = \sigma[P]$$

$$\frac{\sigma[P]}{2 - \sigma[P]} + \frac{\mu}{2}\left(2 - 3\frac{\sigma[P]}{2 - \sigma[P]}\right) = \sigma[P], \tag{5.9}$$

solving for $\sigma[P]$:

$$\sigma[P] = \frac{1 + 2.5\mu - \sqrt{6.25\mu^2 - 3\mu + 1}}{2}, \tag{5.10}$$

solving for μ:

$$\mu = \frac{\sigma[P] - \sigma[P]^2}{2 - \frac{5}{2}\sigma[P]}. \tag{5.11}$$

Similarly, we derive for niche mutation

$$\sigma[A] + \Delta_{mn}(\sigma[A]) = \sigma[P]$$

$$\frac{\sigma[P]}{2 - \sigma[P]} + \mu\left(1 - \frac{2\sigma[P]}{2 - \sigma[P]}\right) = \sigma[P], \tag{5.12}$$

solving for $\sigma[P]$:

$$\sigma[P] = \frac{1 + 3\mu - \sqrt{9\mu^2 - 2\mu + 1}}{2}, \tag{5.13}$$

solving for μ:

$$\mu = \frac{\sigma[P] - \sigma[P]^2}{2 - 3\sigma[P]}. \tag{5.14}$$

The above equations enable us to determine the expected specificity in the population given a fixed mutation probability. On the other hand, given a desired specificity in the population, we can determine the mutation probability necessary to achieve that specificity.

Table 5.1 shows the resulting specificities in theory and empirically determined in a random function (that is, a Boolean function that returns uniformly randomly either zero or one for each problem instance) and a constant function (that is, a Boolean function that returns always one). Evaluated are niche mutation and free mutation. Empirical runs were carried through using either proportionate selection or tournament selection, which is introduced in Section 5.2.3. In the random function, the empirical results show slightly higher values than predicted by the theory. We note that the resulting specificity values can be roughly approximated by twice the value of mutation. The higher specificity values are mainly due to offspring initialization effects in conjunction with the application frequency of classifiers. In the case of low mutation rates, the fitness decrease in offspring classifiers is quickly overruled by higher relative accuracies. In the case of high mutation rates, the fitness decrease causes lower consequent specificities. These effects are enhanced by the much stronger pressure induced by tournament selection. Figure 5.6 shows the convergence values of the table in graphical form.

Table 5.2 shows necessary mutation rates for desired specificities in theory and derived from experiments. Due the specialization effects caused by offspring initialization, the mutation rate needs to be set slightly lower than

Table 5.1. Converged specificities from empirical results in a constant and a random function with free mutation and niche mutation with proportionate selection and tournament selection as well as in theory.

| | Constant Function | | | | Random Function | | | | Theory | |
| | Free Mu. | | Niche Mu. | | Free Mu. | | Niche Mu. | | | |
μ	Prop.S.	Tour.S.	Prop.S.	Tour.S.	Prop.S.	Tour.S.	Prop.S.	Tour.S.	Free Mu.	Niche Mu.
0.01	0.02091	0.10846	0.02098	0.10773	0.03396	0.16846	0.03276	0.16456	0.01990	0.01980
0.02	0.04034	0.11512	0.04022	0.11358	0.06615	0.19667	0.06447	0.18864	0.03959	0.03918
0.04	0.07846	0.12953	0.07720	0.12641	0.12856	0.23261	0.12300	0.21903	0.07830	0.07668
0.06	0.11418	0.14594	0.11119	0.14076	0.18233	0.25920	0.17180	0.24384	0.11606	0.11240
0.08	0.15109	0.16560	0.14509	0.15790	0.22668	0.27971	0.21360	0.26364	0.15279	0.14629
0.10	0.18357	0.18712	0.17554	0.17743	0.25995	0.29965	0.24340	0.28007	0.18839	0.17830
0.12	0.21762	0.21116	0.20539	0.19864	0.28872	0.31810	0.26615	0.29549	0.22280	0.20841
0.14	0.24856	0.23706	0.23055	0.22128	0.31172	0.33538	0.28641	0.30855	0.25592	0.23661
0.16	0.28025	0.26089	0.25655	0.24380	0.33549	0.35106	0.30432	0.32357	0.28769	0.26293
0.18	0.30818	0.28485	0.27956	0.26548	0.35761	0.36812	0.32255	0.33633	0.31803	0.28740
0.20	0.33648	0.30964	0.30244	0.28554	0.37750	0.38494	0.33820	0.34983	0.34689	0.31010

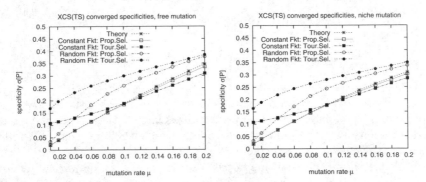

Fig. 5.6. Specificity is strongly dependent on mutation rate and mutation type. In a constant function, the theory nearly matches the resulting specificities. Fitness differences due to offspring initialization and frequency of evaluation change the specificity distribution, especially when tournament selection is applied.

suggested by the theory. Tournament selection enhances this effect since it focuses selection pressure, considering even small fitness differences significant. Note that apart from its effect on specificity, if mutation probability is set too high, mutation may be disruptive, destroying important parental structural information. The result would be an undesired random problem search.

5.2 Improving Fitness Pressure

The previous section has shown that XCS has an intrinsic generalization pressure that needs to be overcome by a sufficiently strong fitness pressure in order to evolve accurate classifiers. Selection in XCS and other LCSs has always

Table 5.2. Mutation settings for desired specificities. Empirical results are derived from simulations in a random function, which returns zero or one equally probable for each problem instance.

		0.05	0.1	0.15	0.2	0.25	0.3	0.4	0.5	0.6
	theory	0.0253	0.0514	0.0785	0.1067	0.1364	0.1680	0.2400	0.3333	0.4800
free mut.	prop.sel.	0.02	0.03	0.05	0.07	.10	0.14	> .2	> .2	> .2
	tour.sel.	< 0.01	< 0.01	0.01	0.02	0.06	0.10	> .2	> .2	> .2
	theory	0.0257	0.0529	0.0823	0.1143	0.1500	0.1909	0.3000	0.5000	–
niche mut.	prop.sel.	0.02	0.03	0.05	0.08	0.11	0.16	> .2	> .2	> .2
	tour.sel.	< 0.01	< 0.01	0.01	0.03	0.07	0.14	> .2	> .2	> .2

been done by means of proportionate selection. As we have seen in Chapter 2, though, proportionate selection results in a fitness pressure that is strongly dependent on fitness scaling and the fitness distribution in the population (Baker, 1985; Goldberg & Deb, 1991; Goldberg & Sastry, 2001; Goldberg, 2002).

This section shows that XCS can actually suffer from the pitfalls of proportionate selection observed in the GA literature. Moreover, we show that tournament selection with tournament sizes proportional to the action set size can solve the problem resulting in a strong, stable, and reliable fitness pressure towards more accurate classifiers.

5.2.1 Proportionate vs. Tournament Selection

Proportionate selection was applied and analyzed in Holland's original GA work (Holland, 1975). However, proportionate selection strongly depends both on fitness scaling (Baker, 1985; Goldberg & Deb, 1991) as well as on the current fitness distribution in the population. The smaller the fitness differences in the population the smaller the fitness pressure. Goldberg and Sastry (2001) show that evolutionary progress stalls when a population comes close to convergence since the fitness differences are not sufficiently strong anymore.

Fitness of XCS classifiers is derived from the scaled, set-relative accuracy. Although fitness scaling usually works well and proportionate selection is applied in the current action sets and not in the whole population, the more similarly accurate classifiers are, the less fitness pressure due to proportionate selection is expectable. In effect, similar accuracy of all classifiers in an action set should decrease or even annihilate fitness pressure.

Tournament selection, on the other hand, does not care about the current relative fitness differences. What matters is fitness rank. Thus, tournament selection does not suffer from fitness scaling nor from very small differences in accuracy. As long as there are significant differences in accuracy, tournament selection detects them and propagates the higher accurate classifier.

The next section shows that proportionate selection does not only suffer from cases in which classifiers are expected to have similar fitness values, but fitness pressure can actually be insufficiently strong. We propose set-size relative tournament selection as the remedy and confirm its superior performance in the exemplar multiplexer problem.

5.2.2 Limitations of Proportionate Selection

To reveal the limitations of proportionate selection, we apply XCS to the multiplexer problem with various parameter settings or with additional noise in the problem. We show that learning in XCS with proportionate selection is disrupted if the learning parameter β is set too low or if problem noise is set too high. The multiplexer problem is introduced in Appendix C.[1]

Figure 5.7 reveals the strong dependence on parameter β. Decreasing the learning rate hinders XCS from evolving an accurate problem solution. The problem is that initially overgeneral classifiers occupy a big part of the population. Better offspring often lose against the overgeneral parents since the fitness of the offspring only increases slowly (due to the low β value). Small differences in the fitness F only have small effects when using proportionate selection. Altering the slope of the accuracy curve by changing parameters α and ε_0 does not have any positive learning effect.

Figure 5.8 reveals XCS's dependence on initial specificity. Increasing $P_\#$ (effectively decreasing initial specificity) impairs the learning speed of XCS, since fitness does not cause sufficient specialization pressure. Decreasing the mutation rate μ also has a detrimental effect strongly delaying learning progress. The generalizing set pressure appears to be often stronger than the specializing fitness pressure so that the chances of reaching higher accurate classifiers are significantly decreased.

Additionally, we detected significant parameter initialization effects when using proportionate selection. If prediction p and reward prediction error ε are set to the parental values, learning speed is slightly decreased. Since in our problems only zero or thousand reward is possible and reward prediction is set directly to either one of the values, if the classifier is accurate, its error will decrease faster and fitness will increase faster.

In addition to the dependency on parameters β, $P_\#$, μ, and initialization, we can show that XCS with proportionate selection is often not able to solve noisy problems. We added two kinds of noise to the multiplexer problem: (1) Gaussian noise with a standard deviation σ is added to the payoff provided by the environment; (2) The payoff is swapped with a certain probability, termed *alternating noise* in the remainder of this work. Figures 5.9 and 5.10 show that XCS's performance is strongly degraded when adding only a small

---

[1] Unless stated otherwise, all results in this section are averaged over 50 experimental runs. Performance is assessed by test trials in which no learning takes place and the better prediction array value is chosen as the classification. During learning, classifications are chosen at random. Parameters are set as follows: $N = 2000$, $\beta = 0.2$, $\alpha = 1$, $\varepsilon_0 = .001$, $\nu = 5$, $\theta_{GA} = 25$, $\chi = 1.0$, $\mu = 0.04$, $\theta_{del} = 20$, $\delta = 0.1$, $\theta_{sub} = 20$, and $P_\# = 0.6$.

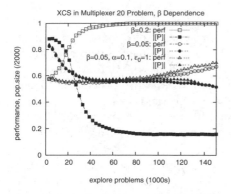

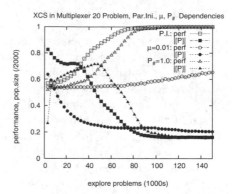

Fig. 5.7. A lower learning rate β decreases XCS learning performance. Accuracy function parameters have no immediate positive influence.

Fig. 5.8. Proportionate selection is strongly dependent on offspring initialization, initial specificity, and mutation rate.

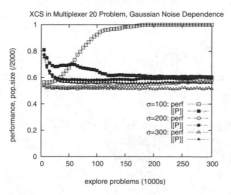

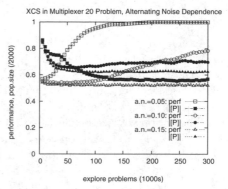

Fig. 5.9. Adding Gaussian noise to the payoff function of the multiplexer problem significantly deteriorates performance of XCS.

Fig. 5.10. Also alternating noise, in which a percentage of examples is assigned to the incorrect class, significantly degrades performance of XCS.

amount of either noise. Similar observations were made in Kovacs (2003), where Gaussian noise was added to the reward in some of the output classes.

In general, the more noise is added, the smaller the fitness difference between accurate and inaccurate classifiers. Thus, selection pressure decreases due to proportionate selection and the population starts to drift at random. Lanzi (1999c) proposed an extension to XCS that detects noise in environments and adjusts the error estimates accordingly. This approach, however, does not solve the parameter dependencies nor problems in which noise is not equally distributed over the problem space.

5.2.3 Tournament Selection

In contrast to proportionate selection, tournament selection is independent of fitness scaling (Goldberg & Deb, 1991). In tournament selection parental classifiers are not selected proportional to their fitness, but tournaments are held in which the classifier with the highest fitness wins (stochastic tournaments are not considered herein). Participants for the tournament are usually chosen at random from the population in which selection is applied. The size of the tournament controls the selection pressure. Fixed tournament sizes are generally used in GAs.

Compared to standard GAs, the GA in XCS is a steady-state, niche GA. Only two classifiers are selected in each GA application and selection is restricted to the classifiers in the current action set. Thus, some classifiers might not get any reproductive opportunity at all before being deleted from the population. Additionally, action set sizes can vary significantly. Initially, action sets are often over-populated with overgeneral classifiers. Thus, a relatively strong selection pressure, which adapts to the current action set size, appears to be necessary.

Thus, effective tournament selection in XCS holds tournaments of sizes dependent on the current action set size $|[A]|$ choosing a subset of size $\tau|[A]|$ ($\tau \in (0, 1]$) of the classifiers in $[A]$.[2] Instead of proportionate selection, two independent tournaments are held in which the classifier with the highest fitness is selected. The tournament selection procedure is described in algorithmic form in Appendix B.

The action set size proportionate tournament size assures that the current best classifier (assuming only one copy) is selected at least once with probability $1 - (1 - \tau)^2$. For example, if the tournament size is set to $\tau = 0.4$ of the population, we assure that the maximally accurate classifier is part of at least one of the two tournaments with a probability of 0.64. On average, $2\tau^2 + 2(\tau(1 - \tau)) = 2\tau$ optimal classifiers are selected. The more copies of the best classifier exist, the higher the probability. This derivation is impossible when fixed tournament sizes are used since the action set size continuously varies and consequently the probability of selecting the best classifier continuously varies as well.

XCSTS in the Previous 20 Multiplexer Settings

Figures 5.11 and 5.12 show that XCS with tournament selection, referred to as *XCSTS*, can solve the 20 multiplexer problem even with a low parameter value β, a low parameter value μ, or a high parameter value $P_\#$. XCSTS is also more independent from initial parameter settings. The much stronger and stable fitness pressure overcomes the generalizing set pressure even without the help of mutation pressure. A reliable and stable performance increase is observable.

---

[2] If not stated differently, τ is set to 0.4 in the subsequent experimental runs.

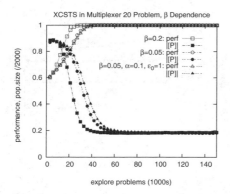

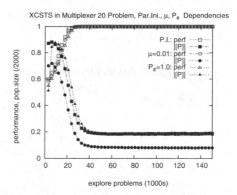

Fig. 5.11. XCSTS is only marginally influenced by a decrease in learning rate β. As in the proportionate selection case, accuracy parameters have nearly no learning influence.

Fig. 5.12. Other parameter variations such as a lower initial specificity ($P_\# = 1.0$), a low mutation rate ($\mu = 0.01$), or a different offspring initialization ($P.I.$) hardly influences XCSTS's learning behavior.

Figures 5.13 and 5.14 show that XCSTS is much more robust in noisy problems as well. XCSTS solves the same Gaussian noise 20 multiplexer problem with nearly no performance degradation (Figure 5.13). Despite the noisy parameter estimation values, tournament selection detects the more accurate classifiers generating a sufficiently strong fitness pressure. The decrease in the differences of accuracy due to the additional noise hardly affects XCSTS. Also in the alternating noise case, XCSTS reaches a higher performance level (Figure 5.14). Note that as expected, the population sizes do not converge to the sizes achieved without noise, since subsumption does not apply. Nonetheless, in both noise cases the population sizes decrease indicating the detection and convergence to maximally accurate classifiers.

Tournament Selection with Fixed Size

XCS's action sets vary in size and in distribution. Dependent on the initial specificity in the population (controlled by parameter $P_\#$), the average action set size is either large or small initially. As was shown above, the average specificity in an action set is on average smaller than the specificity in the whole population. Replication in action sets and deletion from the whole population results in an implicit generalization pressure that can only be overcome by a sufficiently large specialization pressure. Additionally, the distribution of the specificities depends on initial specificity, problem properties, the resulting fitness pressure, and learning dynamics. Thus, an approach with fixed tournament size is dependent on the particular problem and probably not flexible enough. Kovacs used fixed tournament sizes to increase fitness pressure in his comparison to a strength-based XCS version (Kovacs, 2003).

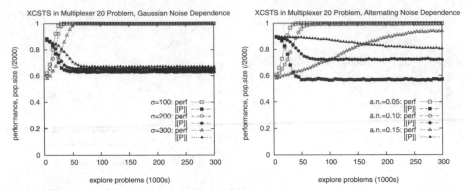

Fig. 5.13. XCSTS performs much better than XCS in noisy problems. Performance is hardly influenced when adding Gaussian noise of up to $\sigma = 300$.

Fig. 5.14. Also in the alternating noise case, XCSTS's learning behavior is faster and more reliable.

In Figure 5.15 we show that XCSTS with fixed tournament size τ_f only solves the multiplexer problem with the large tournament size of $\tau_f = 12$. Since the population is usually over-populated with overgeneral classifiers early in a run, action set sizes are large so that a small tournament size mainly causes competition only among overgeneral classifiers. Thus, not enough fitness pressure is generated. When adding noise, an even larger tournament size is necessary (Figure 5.16). A tournament size of $\tau_f = 32$, however, does not allow any useful recombinatory events anymore since the action set size itself is usually not much bigger than that. Thus, fixed tournament sizes are inappropriate for XCS's selection mechanism.

Different Tournament Sizes

A change in the relative tournament size effectively changes the strength of the selection pressure applied. While a tournament size $\tau = 0$ corresponds to random selection, $\tau = 1$ corresponds to a deterministic selection of the classifier with the currently highest fitness in the action set, and thus the strongest selection pressure. As can be seen in Figure 5.17 and Figure 5.18, XCSTS is able to generate a complete and accurate problem representation for a large range of τ. However, if selection pressure is too weak, learning may not take place at all or may be delayed. On the other hand, if selection pressure is very strong, crossover never has any effect since identical classifiers are crossed. The lack of effective recombinations hardly influences XCS performance in the multiplexer problem as shown in Figure 5.18. However, in other problems ineffective crossover may strongly impair XCS's learning capabilities. Thus, τ may not be set to 1. In practice, a value of 0.4 proved to be robust.

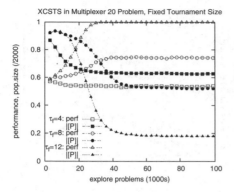

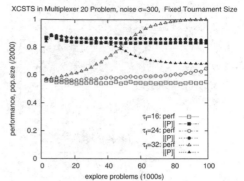

Fig. 5.15. Due to fluctuations in action set sizes and distributions as well as the higher proportion of more general classifiers in action sets, fixed tournament sizes are inappropriate for XCS selection.

Fig. 5.16. Tournament selection with fixed tournament size requires large sizes to solve noisy problems. Large action sets prevent sufficiently strong selection pressure when tournament sizes are too small.

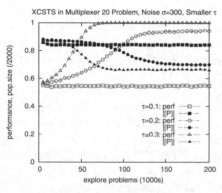

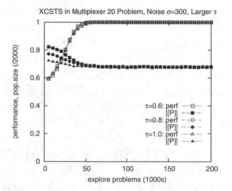

Fig. 5.17. The strength of the selection pressure can be manipulated by the tournament size proportion. Very small proportions result in an insufficiently strong pressure.

Fig. 5.18. Very large tournament set size proportions may prevent effective recombination. In the multiplexer problem, this restriction results in hardly any performance influence.

Specificity Guidance Exhibited

As theorized above, the two selection methods differ in their dependence on the fitness estimate and fitness distribution. The fitness pressure, resulting from proportionate selection, depends on fitness scaling and in particular on the relative amount of fitness difference. Tournament selection on the other hand only depends on the fitness difference itself and not on the amount of the difference. That is, as long as the more accurate classifiers have a higher

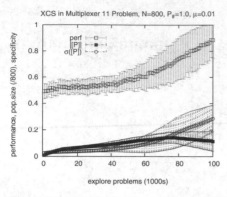 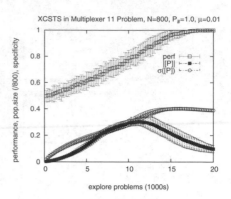

Fig. 5.19. Starting with an overgeneral population and low mutation rate, XCS is not able to pick up the accuracy signal reliably.

Fig. 5.20. XCSTS immediately pushes the population towards higher specificity and thus higher accuracy.

fitness estimate (regardless of how much higher the estimate is), tournament selection causes fitness pressure towards higher accuracy.

To exhibit this pressure, we monitor the average specificity in the population as well as the average standard deviation of the specificity. Figures 5.19 and 5.20 show the change in specificity when starting with a completely general population ($P_\# = 1.0$) in the 11 multiplexer problem. Figure 5.20 shows that XCSTS immediately identifies higher accurate classifies causing the specificity to rise. XCS, on the other hand, stalls at the overgeneral level apparently relying on a lucky guess for successful learning (Figure 5.19).

Another indicator can be found in the performance increase showing also the standard deviation over the experiments. Since XCSTS detects the fitness guidance in the problem immediately, the standard deviation between the runs remains small and the performance is hardly affected by a higher initial generality. Performance of XCS with proportionate selection, on the other hand, is much less reliable and strongly depends on initial generality. If there is no accurate classifier generated initially, proportionate selection has problems generating a sufficiently strong fitness guidance.

5.3 Summary and Conclusions

This chapter has shown how XCS evolves a complete, accurate, and maximally general problem solution. Several *evolutionary pressures* guide the initial classifier population to the solution.

1. *Fitness pressures* is the main pressure towards higher accuracy.
2. *Set pressure* causes classifier generalization towards higher *semantic* generality.

3. *Mutation pressure* causes diversification searching in the syntactic neighborhood of currently best subsolutions. Mutation also has an effect on specificity.
4. *Deletion pressure* emphasizes the maintenance of a complete solution.
5. *Subsumption pressure* propagates accurate classifiers that are *syntactically* more general.

Combining set pressure and mutation pressure, we derived a *specificity equation*, which is able to predict the expected specificity change in an XCS population as long as no additional fitness pressure applies. In conjunction with fitness pressure and subsumption pressure, the pressures push towards an equilibrium that coincides with the desired complete, maximally accurate, and maximally general problem solution.

However, to reach this equilibrium, fitness pressure needs to be strong enough to reliably overcome the generalizing set pressure. We showed that proportionate selection may not be sufficiently strong due to parameter influences as well as problem influences such as a noisy reward function. *Tournament selection* with tournament sizes proportional to the current action set size solves these drawbacks. Since the tournament size is set proportionate to the current action set size, the minimal probability of selecting the most accurate classifier is fixed so that the minimal fitness pressure is also fixed. We showed that XCS with tournament selection is able to solve the investigated problems reliably, confirming the theorized reliable fitness pressure towards higher accuracy.

Note however that tournament selection for deletion is actually inappropriate. Tournament selection strives to evolve the best classifiers and converges to these classifier fast. This is appropriate for reproduction because only one maximally accurate classifier or a few accurate, overlapping classifiers should evolve in each action set. Since deletion is applied population wide and XCS evolves a distributed problem solution, proportionate selection is the right choice in this case since it naturally maintains a distributed problem solution. Horn, Goldberg, and Deb (1994) provides a detailed analysis on the suitability of population wide proportionate selection with fitness sharing.

With respect to the facetwise theory approach for LCSs, we can now assure the first major aspects of the approach: (1) Due to the addition of tournament selection, fitness guides reliably to the intended solution. The accuracy-based fitness approach prevents strong overgenerals; (2) Parameters are estimated appropriately using adapted Q-learning and the moyenne adaptative modifiée technique; (3) Appropriate generalization applies as quantitatively analyzed in the *specificity equation*. Other influences with respect to generalization cause (crossover) or slight additional specificity influence (deletion, parameter initialization, accuracy determination) but no disruption. Subsumption pushes towards maximally syntactically general, accurate classifiers as long as complete accuracy can be reached in the problem (reward prediction error ε drops below ε_0).

With the first main aspect of our facetwise LCS theory understood and satisfied, we are now ready to face the second aspect. The next chapter consequently investigates the computational effort necessary to ensure solution growth and sustenance.

6

When XCS Works:
Towards Computational Complexity

The last chapter investigated fitness guidance and generalization in the XCS classifier system. We saw that several evolutionary pressures apply that are designed to push the evolutionary process towards the desired complete, maximally accurate, and maximally general problem solutions. To ensure reliable fitness pressure, offspring should be selected using *tournament selection* with tournament sizes proportional to the current action set size to ensure a constant minimal pressure towards selecting better offspring in action sets.

The analysis in the last chapter enables the investigation of the next four points of the facetwise LCS theory approach to ensure solution growth and sustenance:

- *Population initialization* needs to ensure a sufficiently general population to provide enough time for classifier evaluation and successful GA application.
- *Schema supply* needs to be ensured to have better classifiers available.
- *Schema growth* needs to be ensured to grow those better classifiers evolving a complete problem solution.
- *Solution sustenance* needs to be ensured to sustain a complete problem solution.

We address these issues in the subsequent sections.

First, we derive a *covering bound* to ensure proper initialization. Second, we derive a *schema!bound* to ensure the availability of better schema representatives for reproduction. We use the schema bound to derive initial settings for population specificity and population size. The time it takes to generate a better classifier at random is also considered. Third, we derive the *reproductive opportunity bound* to ensure that better classifiers can be detected and reproduced successfully. We show that the schema bound and the reproductive opportunity bound somewhat interact since, intuitively, too much supply implies too large specificity, consequently disabling reproduction.

While the reproductive opportunity bound assures that better classifier grow, we are also interested in how long it takes them to grow. This is expressed in the learning time bound derived next.

Once enough time is given to make better classifiers grow until the final solution is expected to be found, we finally need to assure *sustenance* of the grown solution. XCS applies niching in that it reproduces in problem subspaces and deletes from the whole population. Larger niches are preferred for deletion. The analysis results in a final population size bound ensuring the sustenance of a complete problem solution as long as there are no severe problem solution overlaps.

Putting the results together, a positive computational learning theory result is derived for XCS with respect to *k-DNF* functions (problems in disjunctive normal form with maximally k literals per conjunction). We show that when restricting the problem space to *k-DNF* functions, XCS is able to learn an approximately correct solution with high probability in polynomial time, that is, XCS is able to PAC-learn (Mitchell, 1997) *k-DNF* functions with few additional constraints. However, the reader should keep in mind that XCS is a system that is much more broadly applicable and actually an online generalizing, evolutionary RL system. Nonetheless, XCS's capability of PAC-learning *k-DNF* functions confirms its general learning scalability and potential widespread applicability.

6.1 Proper Population Initialization:
The Covering Bound

Several issues need to be considered when intending to make time for classifier evaluation and thus the identification of better classifiers. The first bound is derived from the rather straight-forward requirement that the evolutionary algorithm needs to apply. That is, reproduction in action sets and deletion in the population needs to take place.

The requirement is not met if the initial population is over-specialized since this can cause XCS to get stuck in an infinite covering-random deletion cycle. Normally, covering occurs only briefly in the beginning of a run. However, if covering continues indefinitely because inputs are continuously not covered due to an over-specialized population, the GA cannot take place and the evolutionary pressures do not apply.

As described in the XCS introduction, covering creates classifiers given a problem instance that is not matched by at least one classifier for each possible classification. Since the population is of fixed size N, if the population is already filled up with classifiers, other classifiers are deleted to make room for the newly created covering classifiers. Early in a learning run classifiers in the population will have been evaluated hardly at all so that the fitness value F as well as the action set size estimate as of these classifiers will be basically meaningless. Thus, the deletion method will basically choose classifiers for deletion at random early in a run.

To initiate learning effectively, XCS may not get stuck in such a *covering-random deletion* cycle, in which covering is triggered continuously due to an overspecialized classifier population and deletion is consequently basically random. The initial average specificity in the population depends on parameter $P_\#$, which specifies the probability of a don't care symbol in a classifier condition. If the initial average specificity in the population is too high, the problem space may not be completely covered and the *covering-random deletion* cycle may occur. The covering bound prevents this cycle by bounding population size and initial specificity to assure proper learning initiation.

Assuming a uniform problem instance distribution over the whole problem space $\mathcal{S} = \{0, 1\}^l$, we can determine the probability that a given problem instance is covered by at least one classifier in a randomly generated population:

$$P(\text{cover}) = 1 - \left(1 - \left(\frac{2 - \sigma[P]}{2}\right)^l\right)^N, \tag{6.1}$$

where $\sigma[P]$ may be equated with $1 - P_\#$ since the covering bound only applies in the beginning of a run. Similarly, we can derive the maximal specificity given a certain population size using the inequality $1 - \exp^{-x} < x$ setting $(1 - P(\text{cover}))$ to $1/\exp$:

$$\sigma[P] < 2(1 - (\frac{1}{N})^{1/l}) < 2 - 2(1 - (1 - \text{cover})^{1/N})^{1/l}, \tag{6.2}$$

which shows that increasing N results in an increase in maximal specificity polynomial in $1/l$ deriving an effective rule of thumb of how low the don't care probability $P_\#$ may be set to assure an appropriate specificity $\sigma[P]$. Figure 6.1 shows the resulting boundary conditions on population size and specificity for different problem lengths requiring a confidence level of 0.01.

To automatically avoid the covering bound, XCS could be enhanced to detect infinite covering and consequently increase the $P_\#$ value. However, we did not experiment with such an enhancement so far, since usually the covering bound can be easily circumvented by setting the parameter $P_\#$ large enough.

Given that the problem space is not sampled uniformly at random, the covering bound can be used as an upper bound. The smaller the set of sampled problem instances, the larger the probability that an instance is covered, since the covering mechanism generates classifiers that cover actual instances and the genetic algorithm mainly focuses on generating offspring classifiers that apply in the current problem niche. Since in most RL problems as well as in datamining problems the number of distinct problem instances is usually much smaller than the whole problem space, the covering bound becomes less important in these cases.

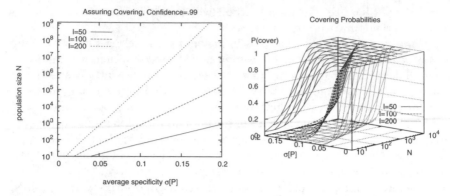

Fig. 6.1. To ensure that better classifiers can be identified, population size needs to be set high enough and specificity low enough to satisfy the covering bound.

6.2 Ensuring Supply: The Schema Bound

The supply question relates to the schema supply in GAs. GAs process BBs. However, BBs may be misleading, as shown in the introduction of the trap problem in Chapter 2. While the fitness structure of one BB may point towards the mediocre solution (that is, a local optimum), the optimal solution to a BB as a whole needs to be processed. Thus, disregarding mutation effects, optimal BB structure needs to be available from the beginning.

The same observation applies in XCS, albeit in slightly different form. The question then arises—what is a BB in the XCS classifier system? We know that fitness is based on accuracy. Thus, a BB should be a substructure in the problem that increases classification accuracy. As there are BBs for GAs, there are minimal substructures in classification problems that result in higher accuracy (and thus fitness).

To establish a general notion of BBs in XCS, we use the notion of a *schema* as suggested elsewhere (Holland, 1971; Holland, 1975; Holland, 1995). A schema for an input of length l is defined as a string that specifies some of the positions and ignores others. The number of specified positions is termed the *order* k of the schema. A schema is said to be *represented* by a classifier if the classifier correctly specifies *at least* all positions that are specified in the schema. Thus, a representative of a schema of order k has a specificity of at least $\sigma(.) = k/l$. For example, a classifier with condition C =##10#0 is a representative of schema **10*0, but also of schemata **10**, **1**0, ***0*0, **1***, ***0**, *****0, and ******.

Let's consider now a specific problem in which a minimal order of at least k_m bits need to be specified to reach higher accuracy. We call such a problem a problem that has a *minimal order of difficulty* k_m. That is, if less than k_m bits are specified in the problem, the class distribution is equal to the overall class distribution. In other words, the *entropy* of the class distribution decreases only if at least some k_m bits are specified. Since XCS's fitness is derived from accuracy, sufficient representatives of the schema of order k_m need to be present in the population to ensure learning success.

6.2.1 Population Size Bound

To assure the supply of the representatives of a schema of order k_m, the population needs to be specific enough and large enough to avoid the covering bound. It is possible to determine the probability that a randomly chosen classifier from the current population is a schema representative by

$$P(\texttt{representative}) = \frac{1}{n}\left(\frac{\sigma[P]}{2}\right)^{k_m}, \tag{6.3}$$

where n denotes the number of possible actions and $\sigma[P]$ denotes the specificity in the population, as before. From Equation 6.3 we can derive the probability of the existence of a representative of a specific schema in the current population:

$$P(\texttt{representative exists}) = 1 - \left(1 - \frac{1}{n}\left(\frac{\sigma[P]}{2}\right)^{k_m}\right)^N, \tag{6.4}$$

basically deriving the probability that at least one schema representative of order k_m exists in $[P]$.

As we noted in the last chapter, in a problem in which no current fitness pressure applies, specificity can be approximated by twice the mutation probability μ, that is, $\sigma[P] \approx 2\mu$. Additionally, the population may be initialized to a desired specificity value by choosing parameter $P_\#$ appropriately. It should be kept in mind, though, that albeit $P_\#$ might bias specificity further early in the run, without any fitness influence, specificity converges to the values derived in the previous chapter and approximated by 2μ. Thus, mutation determines specificity in the long term. Well-chosen $P_\#$ values may boost initial XCS performance.

Requiring a high probability for the existence of a representative, we can derive the following population size bound using the inequality $x < -ln(1-x)$:

$$N > -n\left(\frac{2}{\sigma[P]}\right)^{k_m} ln(1 - P(\texttt{rep.exists})) > \frac{\log\left(1 - P(\texttt{rep.exists})\right)}{\log\left(1 - \frac{1}{n}\left(\frac{\sigma[P]}{2}\right)^{k_m}\right)}, \tag{6.5}$$

which shows that N needs to grow logarithmically in the probability of error, and exponentially in the minimal order of problem difficulty k_m given a certain specificity. Enlarging the specificity, we can satisfy the schema bound. However, schema growth may be violated if specificity is chosen too high as shown in the subsequent sections.

Figure 6.2 shows the schema bound plotting the required population size with respect to a given specificity and several minimal orders of problem

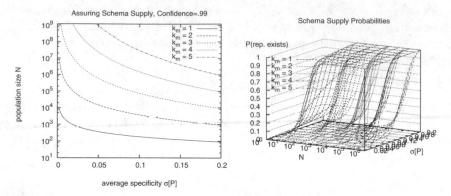

Fig. 6.2. The schema bound requires population size to be set sufficiently high with respect to a given specificity. The larger the population size, the lower the necessary specificity.

difficulty k_m (left-hand side). Population size is plotted in log scale due to its exponential dependence on the order k_m. We also show a 3-D plot of the behavior of the probability that a representative exists, formulated in Equation 6.4.

6.2.2 Specificity Bound

Similar to the bound on population size, given a certain specificity we can derive a minimal specificity bound from Equation 6.4 assuming a fixed population size:

$$\sigma[P] > 2n^{1/k_m}(1 - (1 - P(\texttt{rep.exists}))^{1/N})^{1/k_m}. \tag{6.6}$$

Setting $(1 - P(\texttt{rep.exists}))$ to $1/\exp$ we can use the inequality $1 - \exp^{-x} < x$ to derive that

$$\sigma[P] > 2\left(\frac{n}{N}\right)^{1/k_m}. \tag{6.7}$$

Note that an identical derivation is possible determining the expected number of schema representatives in $[P]$ given specificity $\sigma[P]$:

$$E(representative) = \frac{N}{n}\left(\frac{\sigma[P]}{2}\right)^{k_m}. \tag{6.8}$$

Requiring that at least one representative can be expected in the current population:

$$E(representative) > 1, \tag{6.9}$$

yields again Equation 6.7.

We may rewrite Equation 6.7 using the O-notation. Given a population size of N and the necessary representation of unknown schemata of order k_m, the necessary specificity $\sigma[P]$ can be bounded by

$$\sigma[P] :\ O\left(\left(\frac{n}{N}\right)^{1/k_m}\right),\qquad(6.10)$$

showing that the required specificity decreases polynomially with increasing population size N and increases exponentially in problem complexity k_m. Since we have shown that population size N also needs to increase exponentially in k_m (Equation 6.5) but necessary specificity decreases polynomially in N, the two effects cancel each other out so that it is possible to leave specificity, and thus mutation, untouched, focusing only on a proper population size to assure effective schema supply.

6.2.3 Extension in Time

Given we start with a completely general or highly general initial classifier population (that is, $P_\#$ is close to 1.0), the schema bound also extends in time. In this case, it is the responsibility of mutation to push the population towards the intended specificity generating initial supply.

Given a mutation probability μ, the probability can be approximated that a classifier is generated that has all k_m relevant positions specified given a current specificity $\sigma[P]$:

$$P(\textbf{generation of representative}) = (1 - \mu)^{\sigma[P]k_m} \cdot \mu^{(1-\sigma[P])k_m}. \quad(6.11)$$

With this probability, we can determine the expected number of steps until at least one classifier may have the desired attributes specified. Since this is a geometric distribution,

$$E(t(\textbf{generation of representative})) =$$
$$1/P(\textbf{generation of representative}) =$$
$$\left(\frac{\mu}{1-\mu}\right)^{\sigma[P]k_m} \mu^{-k_m}. \qquad(6.12)$$

Given a current specificity of zero, the expected number of steps until the generation of a representative consequently equals to μ^{-k_m}. Thus, given we start with a completely general population, the expected time until the generation of a first representative is less than μ^{-k_m} (since $\sigma[P]$ increases over time). Requiring that the expected time until a classifier is generated is smaller than some threshold Θ, we can generate a lower bound on the mutation μ as follows:

$$\mu^{-k_m} < \Theta$$
$$\mu > \Theta^{\frac{1}{-k_m}} \qquad(6.13)$$

The extension in time is directly correlated with the specificity bound in Equation 6.7. Setting Θ to N/n we get the same bound (since σ can be approximated by 2μ).

As mentioned before, although supply may be assured easily by setting the specificity σ and thus $P_\#$, and more importantly the mutation rate μ sufficiently high, we yet have to assure that the supplied representatives can grow. This is the concern of the following section in which we also evaluate the schema bound experimentally in conjunction with the derived reproductive opportunity bound.

6.3 Making Time for Growth: The Reproductive Opportunity Bound

To ensure the growth of better classifiers, we need to ensure that *better* classifiers get reproductive opportunities. So far, the covering bound only assures that reproduction and evaluation are taking place. This is a requirement for ensuring growth but not sufficient for it. This section derives and evaluates the *reproductive opportunity!bound* that provides a population size and specificity bound that assures the growth of better classifiers.

The idea is to assure that more accurate classifiers undergo reproductive opportunities before being deleted. To do this, we minimize the probability of a classifier being deleted before reproduced. The constraint effectively results in another population and specificity bound since only a larger population size and a sufficiently small specificity can assure reproduction before deletion.

6.3.1 General Population Size Bound

Let's first determine the expected number of steps until deletion. Assuming neither any fitness estimate influence nor any action set size estimate influence, the probability of deletion is essentially random, as already assumed in Chapter 5. Thus, the probability of deleting a particular classifier in a learning iteration equals

$$P(\texttt{deletion}) = \frac{2}{N}, \tag{6.14}$$

since two classifiers are deleted per iteration. A reproductive opportunity takes place if the classifier is part of an action set. As we have seen in the previous chapter, the introduced tournament selection bounds the probability of reproduction of the best classifier from below by a constant. Thus, the probability of being part of an action set directly determines the probability of reproduction:

$$P(\texttt{reproduction}) = \frac{1}{n} 2^{-l\sigma(cl)}. \tag{6.15}$$

Note that this probability assumes binary input strings, and further that all 2^l possible binary input strings occur with equal probability. Combining Equation 6.14 with Equation 6.15, we can determine that neither reproduction nor deletion occurs at a specific point in time:

$$P(\text{no rep., no del.}) = (1 - P(\text{del.}))(1 - P(\text{rep.})) =$$

$$= (1 - \frac{2}{N})(1 - \frac{2^{-l\sigma[P]}}{n}) = 1 - \frac{2}{N} - \frac{2^{-l\sigma[P]}}{n}(1 - \frac{2}{N}). \qquad (6.16)$$

Together with equations 6.14 and 6.15, we can now derive the probability that a certain classifier is part of an action set before it is deleted:

$$P(\text{rep.before del.}) = P(\text{rep.})(1 - P(\text{del.})) \sum_{i=0}^{\infty} P(\text{no rep., no del.})^i =$$

$$P(\text{rep.})(1 - P(\text{del.})) \frac{1}{1 - P(\text{no rep., no del.})} =$$

$$\frac{\frac{1}{n}2^{-l\sigma[P]}(1 - \frac{2}{N})}{\frac{2}{N} + \frac{1}{n}2^{-l\sigma[P]}(1 - \frac{2}{N})} = \frac{\frac{1}{n}2^{-l\sigma[P]}}{\frac{2}{N-2} + \frac{1}{n}2^{-l\sigma[P]}} = \frac{N-2}{N - 2 + n2^{l\sigma[P]+1}}. \qquad (6.17)$$

Requiring a certain minimal reproduction before deletion probability and solving for the population size N, we get the following bound:

$$N > \frac{2n2^{l\sigma[P]}}{1 - P(\text{rep.before del.})} + 2 \qquad (6.18)$$

This bounds the population size by $O(n2^{l\sigma})$. Since specificity σ can be set proportional to $\sigma = 1/l$, the bound actually usually diminishes. However, in problems in which the problem complexity $k_m > 1$, we have to ensure reproductive opportunities to classifiers that represent order k_m schemata.

The expected specificity of such a representative of an order k_m schema can be estimated given a current specificity $\sigma[P]$. Given that the classifier specifies all k positions, its expected average specificity can be approximated by

$$E(\sigma(\text{repres.of schema of order } k_m)) = \frac{k_m + (l - k_m)\sigma[P]}{l}. \qquad (6.19)$$

Substituting $\sigma(cl)$ from Equation 6.18 with this expected specificity of a representative of a schema of order k, the population size N can be bounded by

$$N > 2 + \frac{n2^{l \cdot \frac{k+(l-k)\sigma[P]}{l}+1}}{1 - P(\text{rep.before del.})}$$

$$N > 2 + \frac{n2^{k+(l-k)\sigma[P]+1}}{1 - P(\text{rep.before del.})}. \qquad (6.20)$$

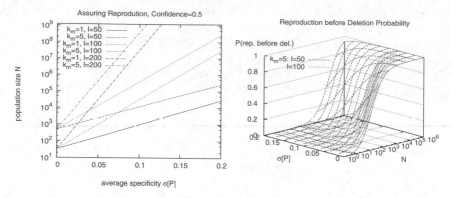

Fig. 6.3. To ensure successful identification and reproduction of better classifiers, population size needs to be set high enough with respect to a given specificity.

This bound ensures that the classifiers necessary in a problem of minimal order of difficulty k_m get reproductive opportunities. Once the bound is satisfied, existing representatives of an order k_m schema are ensured to reproduce before being deleted and XCS is enabled to evolve a more accurate population.

Note that this population size bound is actually exponential in schema order k_m and in string length times specificity $l\sigma[P]$. This would mean that XCS scales exponentially in the problem length which is certainly highly undesirable. However, since specificity in $[P]$ decreases with larger population sizes as shown in Equation 6.10, we show below that a general *reproductive opportunity bound (ROP-bound)* can be derived that shows that population size grows in $O(l^{k_m})$.

Figure 6.3 shows several settings for the reproductive opportunity bound. On the left-hand side, the dependency of population size N on specificity $\sigma[P]$ is shown requiring a confidence value of .99. In comparison to the covering bound, shown in Figure 6.1, it can be seen that the reproductive opportunity bound is always stronger than the covering bound, making the latter nearly obsolete. However, the covering bound is still useful to set the don't care probability $P_\#$ appropriately. The mutation rate, and thus the specificity the population converges to, however, is constrained more strongly by the reproductive opportunity bound. Figure 6.3 also shows a 3-D plot of the dependence of the probability of reproduction before deletion on population size and specificity (right-hand side).

6.3.2 General Reproductive Opportunity Bound

While the above bound ensures the reproduction of *existing* classifiers that represent a particular schema, it does not assure the actual presence or generation of such a classifier. Thus, we need to *combine schema bound* and *reproductive opportunity bound*. Figure 6.4 shows a *control map* of certain re-

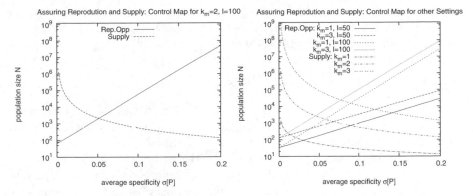

Fig. 6.4. The shown control map clarifies the competition between reproduction and supply. The shown boundaries assure a 50% success probability. High specificity ensures supply but may hinder reproduction. Vice versa, low specificity ensures reproduction but lowers the probability of supply.

productive opportunity bound values and schema bound values requiring a probability of success of 50% (plotting equations 6.5 and 6.20). The corresponding intersections denote the best value of specificity and population size to ensure supply and growth with high probability. Initial specificity may be set slightly larger than the value of the intersection to boost initial performance as long as the covering bound is not violated.

We can also quantify the interaction. Substituting the O-notated specificity bound in Equation 6.10 of the schema bound in the O-notated dependence of the representative bound on string length l ($N : O(2^{l\sigma[P]})$) and ignoring additional constants, we can derive the following enhanced population size bound:

$$N > 2^{l\left(\frac{n}{N}\right)^{1/k_m}}$$

$$\log_2 N > l\left(\frac{n}{N}\right)^{1/k_m}$$

$$N^{1/k_m} \log_2 N > ln^{1/k_m}$$

$$N(\log_2 N)^{k_m} > nl^{k_m}. \tag{6.21}$$

This general *reproductive opportunity bound* (ROP-bound) essentially shows that population size N needs to grow approximately exponentially in the minimal order of problem difficulty k_m and polynomially in the string length.

$$N : O(l^{k_m}) \tag{6.22}$$

Note that in the usual case, k_m is rather small and can often be set to one. When k_m is greater than one, other classification systems in machine learning have also shown similar scale up behavior. For example, the inductive generation of a decision tree in $C4.5$ (Quinlan, 1993) would not be able to decide on which attribute to expand first (since any expansion leads to the same probability distribution and thus no information gain) and consequently would generate an inappropriately large tree.

6.3.3 Sufficiently Accurate Values

Although we generally assume that the estimation values of XCS are sufficiently accurate, we need to note that this assumption does not necessarily hold. All classifier parameters are only an approximation of the average value. The higher parameter β is set, the higher the expected variance of the parameter estimates. Moreover, while the classifiers are younger than $1/\beta$, the estimation values are approximated by the average of the so far encountered values. Thus, if a classifier is younger than $1/\beta$, its parameter variances will be even higher than the one for experienced classifiers.

Requiring that each offspring has the chance to be evaluated at least $1/\beta$ times to get as close to the real value as possible, the reproductive opportunity bound needs to be increased by the number of evaluations we require.

From Equation 6.14 and Equation 6.15, we can derive the expected number of steps until deletion, and similarly, we can derive the expected number of evaluations during a time period t:

$$E(\text{\# steps until deletion}) = \frac{1}{P(\text{deletion})}/2 = \frac{N}{2} \qquad (6.23)$$

$$E(\text{\# of evaluations in } t \text{ steps}) = P(\text{in } [A]) \cdot t = \frac{1}{n}0.5^{l\sigma(cl)}t. \qquad (6.24)$$

The requirement for success can now be determined by requiring that the number of evaluations before deletion must be larger than some threshold Θ where Θ could for example be set to $1/\beta$:

$$E(\text{\# of evaluations in (\# steps until deletion)}) > \Theta$$
$$\frac{1}{n}0.5^{l\sigma(cl)}\frac{N}{2} > \Theta$$
$$N > \Theta n 2^{l\sigma(cl)+1}. \qquad (6.25)$$

Setting Θ to one, Equation 6.25 is basically equal to Equation 6.18 since one evaluation is equal to at least one reproductive opportunity, disregarding the confidence value in this case. It can be seen that the sufficient evaluation bound only increases the reproductive opportunity bound by a constant. Thus, scale-up behavior is not affected.

As a last point in this section, we want to point out that fitness is actually not computed by averaging, but the Widrow-Hoff delta rule is used from the beginning. Moreover, fitness is usually set to 10% of the parental fitness value (to prevent disruption). Thus, fitness is derived from two approximations and it starts off with a disadvantage so that the early estimates of fitness strongly depend on fitness scaling and on accurate approximations of the prediction and prediction error estimates. To ensure a fast detection and reproduction of superior classifiers it is consequently necessary to choose initial classifier values as accurate as possible. Alternatively, the expected variance in fitness values could be considered to prevent potential disruption. Goldberg (1990), for example, suggests using variance sensitive bidding, which incorporates the estimated expectable variance in the selection process. The consequence is that the fitness of young classifiers could be modified to prevent disruption but also to enable the earlier detection of better classifiers.

6.3.4 Bound Verification

This section shows that the derived bounds hold using a Boolean function problem of order of difficulty k_m, where k_m is larger than one. The hidden parity function, introduced in Appendix C and originally investigated in XCS in Kovacs and Kerber (2001), is very suitable to manipulate k_m. The basic problem is represented by a Boolean function in which k relevant bits determine the outcome. If the k bits have an even number of ones, the outcome will be one and zero otherwise. The hidden parity function can also be viewed as an XOR function over k relevant bits in a bit string of length l. The order of difficulty k_m is equivalent to the number of relevant bits k.

The major problem in the hidden parity is that there is no fitness-guidance whatsoever before all k relevant hidden parity bits are specified. That is, if a classifier does not specify the values of all k relevant positions, it has a 50/50 chance of getting the output correct. Thus, the accuracy for all classifiers that do not have all k bits specified is approximately equal.

Figure 6.5 shows that the reproductive opportunity bound is well-approximated by Equation 6.20. The experimental values are derived by determining the population size needed to reliably reach 100% performance. The average specificity in the population is derived from the applied mutation using Table 5.1. XCSTS is assumed to reach 100% performance reliably when all 50 runs reach 100% performance after 200, 000 steps. Although the bound corresponds to the empirical points when specificity is high, in the case of lower specificity the actual population size needed departs from the approximation. The reason for this is the time extension of the schema bound derived in Section 6.2.3 and expressed in Equation 6.12. More detailed evaluations of the reproductive opportunity bound can be found in Butz, Goldberg, and Tharakunnel (2003).

6.4 Estimating Learning Time

Given that schema, covering, and reproductive opportunity bound are satisfied, we essentially ensure the first three aspects of the second main point of our facetwise theory approach: Minimal order conditions are supplied due to

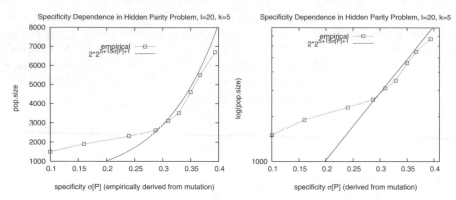

Fig. 6.5. Minimal population size depends on applied specificity in the population (manipulated by varying mutation rates). When mutation and thus specificity is sufficiently low, the bound becomes obsolete.

the schema bound, evaluation time is available due to the covering bound, and finally, time for reproduction is available due to the reproductive opportunity bound. Thus, it is assured that better classifier structures can grow in the population.

Before we address the next aspect in the facetwise theory, we are interested in how long it may take to evolve a complete problem solution by the means of this growing process.

Assuming that the other problem bounds are satisfied by the choice of mutation and population size, we can estimate how long it takes to discover successively better classifiers until the maximally general, accurate classifiers are found. To do this, we assume a domino convergence model (Thierens, Goldberg, & Pereira, 1998) estimating the time until each relevant attribute can be expected to be specialized to the correct value. Considering mutation only, we estimate the time until reproduction and the time until generation of the next best classifier in each problem niche. Using this approach we can show that learning time scales polynomially in problem length and problem complexity.

6.4.1 Time Bound Derivation

To derive our learning time bound, we estimate the time until reproduction of the current best classifier as well as the time until creation of the next best classifier via mutation given a reproductive event of the current best classifier. The model assumes to start with an initially completely general population (that is, $P_\# = 1.0$). Initial specializations are randomly introduced via mutation. Problem-specific initialization techniques or higher initial specificity in the population may speed-up learning time (as long as the covering bound is not violated).

Further assumptions are that the current best classifier is not lost and selected as the offspring when it is part of an action set (assured by the ROP-bound in conjunction with tournament selection). The time model assumes domino convergence (Thierens, Goldberg, & Pereira, 1998) in which each attribute is successively specified. This means that only once the first attribute is correctly specified in a classifier, the second attribute influences fitness and so forth.

Using the above assumptions, we can estimate the probability that mutation correctly specifies the next attribute

$$P(\texttt{perfect mutation}) = \mu(1 - \mu)^{l-1}, \qquad (6.26)$$

where l specifies the number of attributes in a problem instance. This probability can be relaxed in that we only require that the k already correctly set features are not unset (changed to don't care), the next feature is set, and we do not care about the others:

$$P(\texttt{good mutation}) = \mu(1 - \mu)^{k}. \qquad (6.27)$$

Whereas Equation 6.26 specifies the lower bound on the probability that the next best classifier is generated, Equation 6.27 specifies an optimistic bound.

As seen before, the probability of reproduction can be estimated by the probability of occurrence in an action set. The probability of taking part of an action set again, is determined by the current specificity of a classifier. Given a classifier that specifies k attributes, the probability of reproduction is

$$P(\texttt{reproduction}) = \frac{1}{n}\frac{1}{2}^{k}, \qquad (6.28)$$

where n denotes the number of actions in a problem. The best classifier has a minimal specificity of k/l. With respect to the current specificity in the population $\sigma[P]$, the specificity of the best classifier may be expected to be $k + \sigma[P](l - k)$ assuming a uniform specificity distribution in the other $l - k$ attributes. Taking this expected specificity into account, the probability of reproduction is

$$P(\texttt{rep.in [P]}) = \frac{1}{n}\frac{1}{2}^{k+\sigma[P](l-k)}. \qquad (6.29)$$

Since the probability of a successful mutation assumes a reproductive event, the probability of generating an offspring that is better than the current best (that is, that specifies more attributes correctly) is determined by

$$P(\texttt{generation of next best cl.}) = P(\texttt{rep.in [P]})\,P(\texttt{good mutation}) =$$
$$\frac{1}{n}\frac{1}{2}^{k+\sigma[P](l-k)}\mu(1 - \mu)^{l-1}. \qquad (6.30)$$

Since we assume uniform sampling from all possible problem instances, the probability of generating a next best classifier conforms to a geometric distribution (memoryless property, each trial has an independent and equally probable distribution), the expected time until the generation of the next best classifier is

$$E(\texttt{time until gen.of next best cl.}) =$$
$$1/P(\texttt{time until gen.of next best cl.}) =$$
$$\frac{1}{\frac{1}{n}\frac{1}{2}^{k+\sigma[P](l-k)}\mu(1-\mu)^{l-1}} = \frac{n2^{k+\sigma[P](l-k)}}{\mu(1-\mu)^{l-1}} \leq \frac{n2^{k+\sigma[P]l}}{\mu(1-\mu)^{l-1}}. \tag{6.31}$$

Given now a problem in which k_d features need to be specified and given further the domino convergence property in the problem, the expected time until the generation of the next best classifier can be summed to derive the time until the generation of the global best classifier:

$$E(\texttt{time until generation of maximally accurate cl}) =$$
$$\sum_{k=0}^{k_d-1} \frac{n2^{k+\sigma[P]l}}{\mu(1-\mu)^{l-1}} = \frac{n2^{\sigma[P]l}}{\mu(1-\mu)^{l-1}} \sum_{k=0}^{k_d-1} 2^k < \frac{n2^{k_d+\sigma[P]l}}{\mu(1-\mu)^{l-1}}. \tag{6.32}$$

This time bound shows that XCS needs an exponential number of evaluations in the order of problem difficulty k_d. As argued above, the specificity and consequently also mutation needs to be decreased indirect proportional to the string length l. In particular, since specificity $\sigma[P]$ grows as $O((\frac{n}{N})^{\frac{1}{k_m}})$ (Equation 6.10) and population size grows as $O(l^{k_m})$ (Equation 6.22), specificity essentially grows as

$$O(\frac{n}{l}). \tag{6.33}$$

Using the O-notation and substituting in Equation 6.32, we derive the following adjusted time bound making use of the inequality $(1+\frac{n}{l})^l < e^n$:

$$O\left(\frac{l2^{k_d+n}}{(1-\frac{n}{l})^{l-1}}\right) = O\left(\frac{l2^{k_d+n}}{e^{-n}}\right) = O\left(l2^{k_d+n}\right), \tag{6.34}$$

where again parameter n denotes the number of actions. Thus, learning time in XCS is bound mainly by the order of problem difficulty k_d and the number of problem classes n. It is linear in the problem length l. This derivation also validates Wilson's hypothesis that XCS learning time grows polynomially in problem complexity as well as problem length (Wilson, 1998). The next section experimentally validates the derived learning bound.

6.4.2 Experimental Validation

In order to validate the derived bound, we evaluate XCS performance on an artificial problem in which domino convergence is forced to take place. Similar results are expected in typical Boolean function problems in which similar fitness guidance is available, such as in the layered multiplexer problem

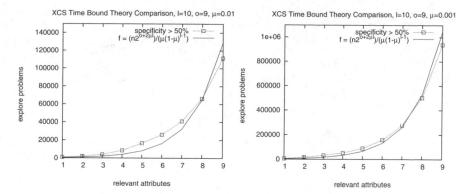

Fig. 6.6. The theory comparison shows that the time until the specificity of the successive attributes has reached 50% is approximated by the theoretical bound. Maximum population size N is set to 32000.

(Wilson, 1995; Butz, Goldberg, & Tharakunnel, 2003). In other problems, additional learning influences may need to be considered such as the influence of crossover or the different fitness guidance in the problem (Butz, Goldberg, & Tharakunnel, 2003). These issues are addressed in more detail in the next chapter.

To validate the time bound, we monitor the specificity of the relevant attributes. According to the domino convergence theory, the system should successively detect the necessary specialization of each relevant attribute, eventually converging to a specificity of nearly 100%. The time bound estimates the expected time until all relevant attributes are detected. To evaluate the bound, we record the number of steps until the specificity of a particular attribute reaches 50%. This criterion indicates that the necessary specificity is correctly detected but it does not require full convergence.

Figure 6.6 shows the time until 50% specificity is reached in the successive attributes in the setting with $l = 10$ and $k_d = 9$. The experimental runs are matching with the theoretical bounds approximating the specificity in the population $\sigma[P]$ with 2μ as done above. As predicted by the theory, decreasing the mutation rate (Figure 6.6, right-hand side) increases the time until the required specificity is reached. Although nearly all interactions between the different niches are prevented by disallowing the mutation of the action part and by applying niche mutation only, the specificity in the later attributes is still learned slightly faster than predicted by the theory. Two-stage interactions might occur in which mutation first overgeneralizes a highly accurate classifier and then specializes it into another niche.

The second concern is the influence of the number of irrelevant attributes. Figure 6.7 shows that also in this case the theory closely matches the empirical results. Since a higher mutation rate results in a higher specificity, the influence of the number of irrelevant attributes is more significant in the setting

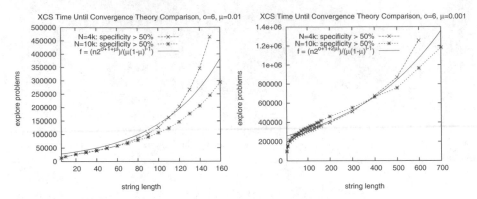

Fig. 6.7. The influence of problem length is properly bound by the theory. The reproductive opportunity bound increasingly outweighs the time bound when the population size is set too low.

with a mutation rate of $\mu = 0.01$. Using a smaller population size can delay or stall the evolutionary process due to the reproductive opportunity bound.

Additional results in Butz, Goldberg, and Lanzi (2004) confirm the successive specialization of relevant attributes when monitoring specificity. It is also shown that free mutation has additional beneficial effects compared to niche mutation, exchanging discovered necessary specializations more effectively. Uniform crossover also contributes to information exchange.

6.5 Assuring Solution Sustenance: The Niche Support Bound

The above bounds assure that problem subsolutions evolve in individual classifiers. The time bound additionally estimates how long the evolution takes. Since the time bound and all other bounds consider individual classifiers integrated in the whole population, the population as a whole is required to evolve a complete problem solution supplying, evaluating, and growing currently best subsolutions in parallel. What remains to be assured is that the final problem solution, represented by a set of maximally accurate and maximally general classifiers, can be sustained in the population. This is expressed in the sixth point of the facetwise theory approach to LCS success: Niching techniques need to assure the sustenance of a complete problem solution.

Thus, this section investigates which niching techniques XCS employs to assure the niche support of all subsolutions. We derive a niche support bound by developing a simple Markov chain model of XCS classifier support. Essentially, we model the change in niche size of particular problem subsolutions (that is, niches) using a Markov chain.

To derive the bound, we focus on the support of one niche only, disregarding potential interactions with other niches. Again we assume that problem instances are encountered according to a uniform distribution over the whole problem space. Additionally, we assume random deletion from a niche. Given the Markov chain over niche sizes, we then determine the steady state distribution that estimates the expected niche distribution.

Using the steady state distribution, we derive the probability that a niche is lost. This probability can be bounded by the population size, resulting in a final population size bound. The bound assures the sustenance of a low-error solution with high probability. The experimental evaluations show that the assumptions hold in non-overlapping problems. In problems that require overlapping solution representations, the population size may need to be increased further.

6.5.1 Markov Chain Model

As already introduced for the schema bound in Section 6.2, we define a problem niche by a schema of order k. A representative of a problem niche is defined as a classifier that specifies at least all k bits correctly. The Markov chain model constructs a Markov chain over the number of classifier representatives in a particular problem niche.

Suppose we have a particular problem niche represented by j classifiers; let p be the probability that an input belonging to the niche is encountered, and let N be the population size. As assumed before, classifiers are deleted approximately at random from the population so that a classifier will be deleted from its niche with probability j/N. Assuming that the GA is always applied (this can be assured by setting $\theta_{GA} = 0$) and disregarding any disruptive effects due to mutation or crossover, the probability that a new classifier is added to the niche is exactly equal to the probability p that an input belonging to the niche is encountered.

However, overgeneral classifiers might inhabit the niche as well so that an overgeneral classifier might also be chosen for reproduction, decreasing the reproduction probability p of a niche representative. As shown in Chapter 5, though, due to the action set size relative tournament selection, the probability of selecting a niche representative for reproduction is larger than some constant dependent on the relative tournament size τ. Given that τ is chosen sufficiently large and given further that the population mainly converged to the niche representatives, the probability approaches one.

In the Markov chain model, we assume that at each time step both GA reproduction and deletion are applied. Accordingly, we derive three transition probabilities for a specific niche. Given that the niche is currently represented by j classifiers, at each time step, (i) with probability r_j the size of the niche is increased (because a classifier has been reproduced from the niche, while another classifier has been deleted from another niche); (ii) with probability d_j the size of the niche is decreased (because genetic reproduction took place

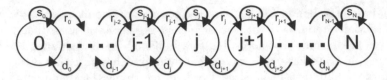

Fig. 6.8. Markov chain model for the support of one niche in XCS: j is the number of classifiers in the niche; N is the population size; r_j is the probability that a classifier is added to a niche containing j representatives; s_j is the probability that the niche containing j representatives is not modified through reproduction; d_j is the probability that a classifier is deleted from a niche containing j representatives.

in another niche, while a classifier was deleted from this niche); (iii) with probability s_j the niche size remains constant (either because no classifier has been added nor deleted from the niche or because one classifier has been added to the niche while another one has been deleted from the same niche).

The Markov chain associated with the model is depicted in Figure 6.8. States in the model indicate the niche size determined by the number of representatives in a niche. Arcs labeled with r_j represent the event that the application of the GA and deletion results in an increase of the niche size. Arcs labeled with s_j represent the event that the application of the genetic algorithm and deletion results in no overall effect on the niche size. Arcs labeled with d_j represent the event that the application of the genetic algorithm and deletion results in a decrease of the niche size.

More formally, since the current problem instance is part of a particular niche with probability p, a niche representative will be generated via GA reproduction with approximate probability p. Assuming random deletion, a representative of a niche is deleted with probability j/N since there are by definition j representatives in the current population of size N. Accordingly, we compute the probabilities r_j, s_j, and d_j as follows:

$$r_j = p \left(1 - \frac{j}{N} \right) \tag{6.35}$$

$$s_j = (1 - p) \left(1 - \frac{j}{N} \right) + p \frac{j}{N} \tag{6.36}$$

$$d_j = (1 - p) \frac{j}{N} \tag{6.37}$$

For $j = 0$ we have $r_0 = p$, $s_0 = 1 - p$, and $d_0 = 0$. When $j = 0$, the niche is not represented in the population, therefore: (i) when an input belonging to the niche is presented to the system (with probability p), one classifier is generated through *covering*, therefore $r_0 = p$; (ii) since the niche has no classifiers, deletion cannot take place, therefore $d_0 = 0$; finally, (iii) the probability that the niche remains unrepresented is $1 - r_0 - s_0$, that is $s_0 = 1 - p$. Similarly, when $j = N$ all the classifiers in the population belong to the niche, accordingly: (i) no classifier can be added to the niche, therefore $r_N = 0$; (ii) with probability

p an input belonging to the niche is encountered so that a classifier from the niche is reproduced while another one from the niche is deleted, leaving the niche size constant, therefore $s_N = p$; (iii) when an input that does not belong to the niche is presented to the system (with probability $1 - p$), a classifier is deleted from the niche to allow the insertion of the new classifier to the other niche, therefore $d_N = 1 - p$. Thus, for $j = N$ we have $r_N = 0$, $s_N = p$, and $d_N = 1 - p$.

Note that our approach somewhat brushes over the problem of overgeneral classifiers in that overgeneral classifiers are not considered as representatives of any niche. In addition, covering may not be sufficient in the event of an empty niche since overgeneral classifiers might still be present so that $r_0 = p$ is an approximation. However, as pointed out by Horn (1993), as long as a sufficiently large population size is chosen, chopping off or approximating the quasi absorbing state r_0 approximates the distribution accurately enough. This is also confirmed by the provided experimental investigations. Given the above assumptions, we are now able to derive a probability distribution over niche support.

6.5.2 Steady State Derivation

To estimate the distribution over the number of representatives of a problem niche, we derive the steady-state distribution over the Markov chain. Essentially, we derive probabilities u_j that the niche has j representatives. Since in steady state the incoming proportion needs to be equal to the outgoing proportion in each state in the Markov chain, the fixed point equation of the Markov chain can be written as follows:

$$(r_j + d_j)u_j = r_{j-1}u_{j-1} + d_{j+1}u_{j+1} \tag{6.38}$$

Replacing d_{j+1}, d_j, r_j, and r_{j+1} with the actual values from the previous section (equations 6.35, 6.36, and 6.37) we get

$$\left[p\left(1 - \frac{j}{N}\right) + \frac{j}{N}(1 - p) \right] u_j =$$

$$= (1 - p)\left(\frac{j+1}{N}\right) u_{j+1} + p\left(1 - \frac{j-1}{N}\right) u_{j-1}. \tag{6.39}$$

Equation 6.39 is a second order difference equation whose parameters are dependent on j, i.e., on the current state. We use Equation 6.39 and the condition:

$$\sum_{j=0}^{j=N} u_j = 1, \tag{6.40}$$

to derive the consequent steady state distribution. The derivation yields:

$$u_j = \binom{N}{j} p^j (1-p)^{N-j}, \tag{6.41}$$

which shows that the interaction of linearly increasing deletion probability with constant reproduction probability results in a binomial distribution. The exact derivation can be found in Butz, Goldberg, Lanzi, and Sastry (2004).

Before we proceed with the derivation of the consequent population size bound, we provide some experimental evaluations investigating the severeness of the made assumptions.

6.5.3 Evaluation of Niche Support Distribution

Our evaluation focuses on two Boolean function problems introduced in Appendix C: (1) the layered count ones problem; and (2) the carry problem. While the layered count ones problem requires a non-overlapping solution representation, the carry problem requires an overlapping solution representation. Apart from the effect of overlapping solutions, we also evaluate the influence of several parameter settings.

Layered Count Ones Problem

In the layered count ones problem, each niche is represented by specializing a subset of k relevant attributes. There are 2^k such non-overlapping subsets. Since we generate problem instances uniformly randomly, the probability p of reproduction in a particular subset equals $p = 1/2^k$. We evaluate the theory on the layered count ones problem with $k = 5, l = 11$. We run XCS for $100,000$ steps to assure that the problem is learned completely. To evaluate the effects of mutation and crossover, we ran additional runs with $50,000$ subsequent condensation steps, in which GA reproduction and deletion are executed as usual but neither mutation nor crossover are applied (Wilson, 1995). Other parameters are set to the standard values introduced earlier. The niche size distribution is derived from all 2^k niche support sizes of each of the 1000 experimental runs.

Applying tournament selection and continuous GA application ($\theta_{GA} = 0$), Figure 6.9 (left-hand side) shows the resulting distributions without condensation varying the population sizes and Figure 6.9 (right-hand side) shows the resulting distributions when an additional $50,000$ condensation steps were executed. Without condensation, the actual distributions are slightly overestimated by the theory, that is, the theory predicts larger niche support. This is explainable by the continuous application of mutation and crossover that causes the reproduction of overgeneral classifiers that are not representing any niche. When condensation is applied and niche-loss is prevented by a sufficiently large population size, the theory agrees nearly exactly with the experiments. If the population size is chosen too low ($N = 500$) niches are continuously lost and cannot be recovered once condensation applies. Thus,

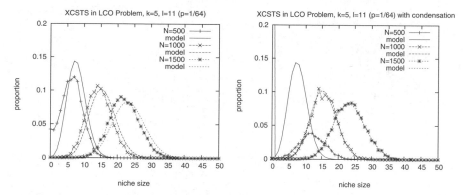

Fig. 6.9. The niche distribution is closely matched by the theory. Eliminating the influences of mutation and crossover results in a nearly exact match.

the niches that were not lost have an on average higher support but a large fraction of niches is lost.

We made several assumptions in the theory including the continuous application of the GA (assured by setting $\theta_{GA} = 0$). To observe the effect of θ_{GA}, we set the parameter to 200 and re-ran the experiments. Figure 6.10 shows the consequent distributions. It can be seen that the distributions have a smaller deviation, effectively focusing on the mean. Due to the smaller deviation, the probability of loosing a niche is significantly decreased as indicated by the maintenance of an appropriate support distribution even with low population size (Figure 6.10, right-hand side, $N = 500$). Given that a problem niche is sampled by several problem instances in succession, the threshold effectively prevents an over-reproduction in the niche by preventing GA reproduction. Thus, the GA threshold prevents the frequent reproduction in one problem niche and consequently causes a further balance in the niche size distribution. In essence, our Markov-chain approximation does not hold anymore when $\theta_{GA} > 0$ because the probability of reproduction in a niche now depends also on the recent history of reproductive events and not only on the probability of niche occurrence. The result is the prevention of niche over-reproduction and consequently the further prevention of niche loss. Thus, the niche size distribution has a smaller variance but the mean stays approximately the same. If niche loss if prevented due to the lower variance, a very significant distribution change is encountered.

Carry Problem

The carry problem is a typical problem in which overlapping classifiers are evolved to generate a complete solution. Additionally, the niche sizes of the overlapping classifiers differ. That is, the probability of a niche occurrence p differ. Our investigations consider the still very small carry-3 problem. We add

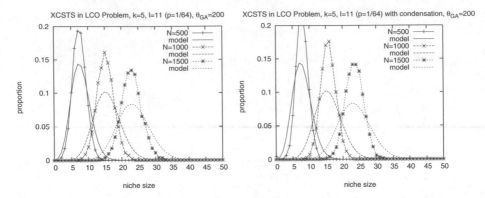

Fig. 6.10. Increasing the GA threshold results in a more balanced distribution causing an overall decrease in deviation from the expected distribution mean.

five irrelevant bits so that the problem is as large as the eleven multiplexer problem. In this problem two binary numbers of length three are added. If the result triggers a carry, then the class is one and zero otherwise. For example, to assure that a carry occurs, it is sufficient to assure that the highest two bits are set to one, that is, 1##1##→1 is a completely accurate and maximally general classifier. However two ones at the second position and at least one one at the first also suffice to assure the event of a carry. That is, classifier 11##1#→1 is also completely accurate and maximally general but it overlaps with the first classifier in the 11*11* subspace. This partial competition also influences the niche distribution.

Figure 6.11 shows the resulting niche distributions of all 36 niches in the problem. Note how the zero niches (left-hand side) have a lower frequency than the correspondingly specific one niches (e.g. 0##0####### vs. 1##1#######) due to the higher overlap with other competitors. Note also how the classifiers that specify three zeros in the two higher bits of the two numbers have a slightly smaller distribution than the other classifiers that specify three zeros. Again, the additional competition due to more overlaps in the former is responsible for the effect.

Due to the overlapping final solution, it is not immediately possible to derive the probability of reproduction for a particular niche. However, it is possible to derive a general probability of reproduction for a subset of niches that are non-overlapping with the other niches. In the carry problem, the zero niches clearly do not overlap with the one niches so that the overall distribution of the zero and one niches can be estimated with our theory. The overall reproductive probability p for all representatives of all one niches can be estimated by summing the probability that the highest bits are both one (0.25), plus the event that both second highest bits but only one of the highest bits are one (0.125), plus the event that both third-highest positions are one and only one of the two higher positions is one (0.0625). In sum, the

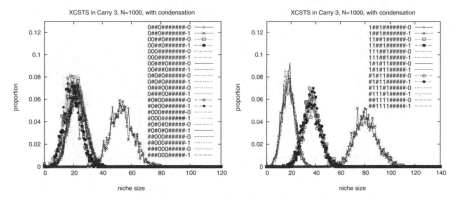

Fig. 6.11. The overlapping, differently specific niches in the carry-3 problem cause interference. This makes it hard to predict the final niche size distribution accurately.

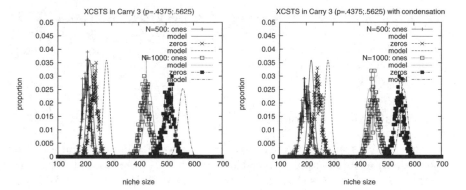

Fig. 6.12. Combining all zero and all one niches and comparing the resulting (macro-)distributions shows that the model applies again. The action set size based deletion in conjunction with the overlapping nature of the problem cause the two distributions to move slightly closer together. As before, condensation eliminates overgeneral classifiers and thus focuses the population on niche representatives .

result equals 0.4375. Figure 6.12 shows the overall distribution. It appears that the action set size based deletion pushes the distribution slightly closer to each other, which is also confirmed in the runs with a higher GA threshold (Figure 6.13).

In sum, we saw that solution spaces interfere with each other during selection in problems that require an overlapping solution representation. The overlap causes a decrease in niche sizes. However, the influence is not as significant as originally feared. Further extensions to balance the niches are imaginable, such as taking into consideration the degree of overlap among competing (fitness sharing) classifiers.

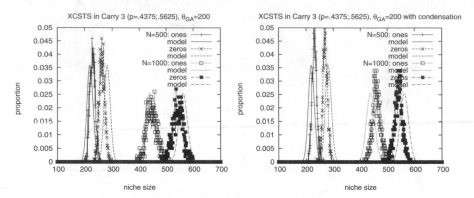

Fig. 6.13. A higher GA threshold moves the lower and upper distributions closer together and slightly focuses the population.

Nonetheless, the model is able to predict the general behavior of XCS's final solution. Additionally, the model can be used to estimate the probability of a niche loss. The next section uses this probability to derive a general population size bound that ensures the sustenance of a low-error solution with high probability.

6.5.4 Population Size Bound

The derived niche support model provides an asymptotic prediction of niche support distribution. It applies once the problem has been learned. The influences of genetic operators showed to cause slight additional niche support disruption, which may require a slightly increased overall population size bound. Besides its predictive capability, the model can be used to derive a population size bound that ensures that a complete model is sustained with high probability.

In particular, using Equation 6.41, a bound on population size N can be derived that ensures with high probability that XCS does not loose any of the problem niches, that is, any subsolutions. From the derivation of the probability of being in state u_0 (which means that the respective niche was lost), which is $u_0 = (1 - p)^N$, we see that the probability of loosing a niche decreases exponentially with the population size. Given a problem with 2^k equally frequently occurring problem niches, that is, the perfect solution $[O]$ is represented by 2^k schemata of order k, the probability of loosing a niche equates

$$u_0 = \left(1 - \frac{1}{2^k}\right)^N. \tag{6.42}$$

Requiring a certainty θ that no niche is lost (that is, $\theta = 1 - u_0$), we can derive a concrete population size bound

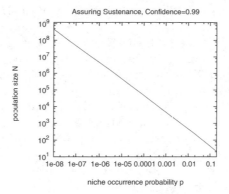

Fig. 6.14. To sustain a complete solution with high probability, population size needs to grow as the inverse of the niche occurrence probability p. In a uniformly sampled, binary problem, p corresponds to $1/2^k$ where k specifies the minimal number of attributes necessary to do maximally accurate predictions in all solution subspaces.

$$N > \frac{\log(1 - \theta)}{\log(1 - p)} > \frac{\log(1 - \theta)}{\log(1 - \frac{1}{2^k})}, \tag{6.43}$$

effectively showing that population size N grows logarithmically in the confidence value and polynomially in the solution complexity 2^k. Figure 6.14 shows the population size bound that assures niche support. Since the population size scales as the inverse of the probability of niche occurrence, the $\log - \log$-sale shows a straight line.

Thus, the bound confirms that once a problem solution was found, XCS is able to maintain the problem solution with high probability requiring a population size that grows polynomially in solution complexity and logarithmically in the confidence value. This bound confirms that XCS does not need more than a polynomial population size with respect to the solution complexity, suggesting the PAC learning capability of XCS.

6.6 Towards PAC Learnability

The derivations of the problem bounds in the previous sections enables us to connect learning in the XCS classifier system to fundamental elements of computational learning theory (COLT). COLT is interested in showing how much computational power an algorithm needs to learn a particular problem. To derive an overall computational estimate of XCS's learning capabilities, we focus on the problem of learning *k-DNF* functions. In particular, we show that *k-DNF* problems that satisfy few additional properties are PAC-learnable (Valiant, 1984; Mitchell, 1997) by XCS. In essence, we also confirm Wilson's previous conjecture that XCS scales polynomially in time and space complexity (Wilson, 1998).

XCS is certainly not the most effective *k-DNF* learning algorithm. Servedio (2001) showed that an algorithm especially targeted to solve noise-free uniformly sampled *k-DNF* problems is able to reach a much more effective performance. However, we do not only show that XCS is able to PAC-learn *k-DNF* problems. We also show that it is able to learn a large variety of problems including nominal and real-valued problems, noisy problems, as well as general RL problems. When restricting the problem to *k-DNF* problems, we can show that XCS is a PAC-learning algorithm confirming the effectiveness as well as the generality of XCS's learning mechanism.

To approach the PAC-learning bound, we reflect on the previous bounds evaluating their impact on computational complexity. The successive chapters provide a variety of evidence for XCS's successful and broad applicability as well as its effective learning and scale-up properties.

6.6.1 Problem Bounds Revisited

In Chapter 5, we analyzed how the evolutionary pressures in XCS bias learning towards the evolution of a complete, maximally accurate, and maximally general problem solution ensuring *fitness guidance* and *appropriate generalization*. This chapter investigated the requirement on population size and learning time in order to supply better classifiers, make time and space to detect and grow those better classifiers until the final population is reached, and finally, to sustain the final problem solution with high probability. Satisfying these bounds, we can ensure with few additional constraints that XCS learns the underlying problem successfully.

We now revisit the bounds considering their resulting computational requirement.

Covering Bound

The covering bound ensures that the GA is taking place establishing a covering probability (Equation 6.1). To ensure a high probability of covering, the specificity can be chosen very low by setting the initial specificity (controlled by $P_\#$) as well as mutation rate sufficiently low. Given a fixed specificity that behaves in $O(\frac{n}{l})$ as necessary to supply better classifiers, as derived above (Equation 6.33), the population size can be bounded as follows using the approximation $x < -\log(1-x)$ and $(1 + \frac{x}{n})^n \approx e^x$:

$$\frac{-\log(1 - P(\text{cov.}))}{-\log\left(1 - \left(\frac{2 - \sigma[P]}{2}\right)^l\right)} < \frac{-\log(1 - P(\text{cov.}))}{\left(1 - \frac{n}{2l}\right)^l} <$$

$$< -\log(1 - P(\text{cov.}))e^{n/2} < N. \tag{6.44}$$

Thus, to satisfy the covering bound, the population size needs to grow logarithmically in the probability of error and exponentially in the number of problem classes n. With respect to PAC-learnability, the bound shows that to ensure that the GA is successfully applied in XCS with probability $1 - \delta_P$ (where $\delta_P = (1 - P(\texttt{cov.}))$), the population size scales logarithmically in the error probability δ_P as well as exponentially in the number of problem classes.

Schema Bound

The *schema bound* ensures that better classifiers are available in the population. Given a problem with a minimal order of difficulty k_m and requiring a high probability that representatives of this order are supplied (Equation 6.3), we were able to bound the population size N in Equation 6.5, showing that population size N needs to grow logarithmically in the probability of error δ_P (that is, $\delta_P = 1 - P(\texttt{rep.exists})$) and exponentially in the minimal order complexity k_m given a certain specificity and thus polynomial in concept space complexity.

Reproductive Opportunity Bound

In addition to the existence of a representative, we showed that it is necessary to ensure reproduction and thus growth of such representatives. This is ensured by the general reproductive opportunity bound, which was shown to require a population size growth of $O(l^{k_m})$ (Equation 6.22) with respect to the minimal order complexity k_m of a problem. The reproductive opportunity bound was generated with respect to one niche. However, since XCS is evolving the niches in parallel, and the probability of niche occurrence as well as the probability of deletion underly a geometric distribution (memoryless property and approximately equal probabilities), we can assure with high confidence $1 - \delta_P$, that all relevant niches receive reproductive opportunities. Thus, we can assure with high probability that lower-order representatives grow leading to higher accuracy within a complexity that is polynomial in concept space complexity.

Time Bound

The time bound estimates the number of problem instances necessary to learn a complete problem solution using XCS. Given a problem that requires classifiers of maximal schema order k_d (a k_d-conjunctive term in a k-*DNF*) and given further the domino convergence property in the problem, the expected time until generation of an optimal classifier was approximated in Equation 6.32 yielding a population size requirement of $O(l2^{k_d+n})$ (Equation 6.34). The estimation approximates the expected time until creation of the best classifier of a problem niche of order k_d. As in the reproductive opportunity bound, we can argue that since XCS evolves all problem niches in parallel and since the generation of the next best classifier underlies a geometric distribution (memoryless property and equal probability), given a certain confidence margin $1 - \delta_P$, the time until a particular classifier of order k is generated with high confidence $1 - \delta_P$ grows within the same limits. Assuming a probability p of problem niche occurrence, we can bound the time until all niches with probability of occurrence $p > \varepsilon_P$ will be discovered with high probability $(1 - \delta_P)$ by $O(l\frac{1}{\varepsilon_P}2^n)$.

Niche Support Bound

To ensure the support of the final solution, we finally established the niche support bound. Given a problem whose solution is expressed as a disjunction of distinct *subsolutions*, XCS tends to allocate distinct rules to each *subsolution*. To ensure a complete problem solution, it needs to be assured that all subsolutions are represented with high probability. Deriving a Markov model over the number of representatives for a particular niche, we were able to derive the steady state niche distribution given a niche occurrence probability p (Equation 6.41).

Requiring that all niches with a niche occurrence probability of at least p are expected to be present in the population with high probability, we were able to derive a bound on the population size N requiring that the probability of no representative in a niche with more than p occurrence probability is sufficiently low. With respect to PAC-learnability this bound requires that with high probability $1 - \delta_P$ we assure that our solution has an error probability of less than ε_P (where ε_P is directly related to p). Using this notation, we can derive the following equation from Equation 6.41 for the distribution of u_0:

$$u_0 = (1 - p)^N, \tag{6.45}$$

substituting ε_P for p and δ_P for u_0, and using again the approximation $x < -\log(1 - x)$ we can derive the following population size bound:

$$\frac{\log \delta_P}{\log(1 - \varepsilon_P)} < -\frac{1}{\varepsilon_P} \log \frac{1}{\delta_P} < N. \tag{6.46}$$

This bound essentially bounds the population size showing that it needs to grow logarithmically in $\frac{1}{\delta_P}$ and linearly in $\frac{1}{\varepsilon_P}$. Approximating ε_P by $(\frac{1}{2})^{k_d}$ assuming a uniform problem instance distribution, we see that to prevent niche loss, the population size needs to grow linearly in the concept space complexity 2^{k_d}.

6.6.2 PAC-Learning with XCS

With the bounds above, we are now able to bound computational effort and number of problem instances necessary to evolve with high probability $(1 - \delta_P)$ a low error ε_P solution of an underlying *k-DNF* problem. Additionally, the *k-DNF* problem needs to be maximally of minimal order of difficulty k_m as discussed in sections 6.2 and 6.3.

Thus, Boolean function problems in *k-DNF* form with l attributes and maximally a minimal order of problem difficulty k_m are PAC-learnable by XCS using the ternary representation of XCS conditions. That is, XCS evolves with high probability $(1-\delta_P)$ a low error ε_P solution of the underlying *k-DNF* problem in time polynomial in $1/\delta_P$, $1/\varepsilon_P$, l, and concept space complexity 2^{k_d}.

The bounds derived in Section 6.6.1 show that the computational complexity of XCS, which is bound by the population size N, is linear in $1/\delta_P$, $1/\varepsilon_P$, and l^{k_m}. Additionally, the time bound shows that the number of problem instances necessary to evolve a low-error solution with high probability grows linearly in $1/\delta_P$, $1/\varepsilon_P$, l and 2^{k_d}. Consequently, we showed that Boolean functions that can be represented in *k-DNF* and have maximally a minimal order of problem difficulty k_m are PAC-learnable by XCS using the ternary representation of XCS conditions as long as the assumptions in the bound derivations hold.

The following further assumptions about the interaction of the bounds have been made. First, crossover is not modeled in our derivation. While crossover can be disruptive, as already proposed by Holland (1975), crossover may also play an important innovative role in recombining currently found subsolutions effectively (Goldberg, 2002). Chapter 7 shows that in XCS crossover can also play the role of the innovator requiring that substructures are effectively processed and recombined.

Second, the specificity derivation from the mutation rate assumes no actual fitness influence. Subtle interactions of various niches might increase specificity further. Thus, problems in which the specificity assumption does not hold might violate the derived reproductive opportunity bound. The experimental investigations in Chapter 8 on several Boolean functions provide further evidence on the influence of specificity and the resulting learning time and population size requirements.

Third, if the probability of reproduction p is approximated by $(\frac{1}{2})^{k_d}$, niche support assumes a non-overlapping representation of the final solution. Thus, overlapping solution representations require an additional increase in population size as evaluated in Section 6.5.

6.7 Summary and Conclusions

This chapter showed *when* XCS is able to learn a problem. Along our facetwise theory approach to LCSs, we derived population size, specificity, and time bounds that assure that a complete, maximally accurate, and maximally general problem solution can evolve and can be sustained.

In particular, we derived a *covering bound* that bounds population size and specificity to ensure proper XCS initialization, making way for classifier evaluation and GA application. Next, we derived a *schema bound* that bounds population size and specificity to ensure *supply* of better classifiers. Better classifiers were defined as classifiers that have higher accuracy on average. They can be characterized as *representatives* of minimal order schemata or BBs—those BBs that increase classification accuracy in the problem at hand. Next, we derived a *reproductive opportunity bound* that bounds specificity and population size to assure *identification* and *growth* of better classifiers. The subsequently derived *time bound* estimates the learning time needed to evolve a complete problem solution given the other bounds are satisfied. Finally, we derived a *niche bound* that assures the *sustenance* of a low-error solution with high probability.

Along the way, we defined two major problem complexities: (1) the minimal order complexity k_m, which specifies the minimal number of features that need to be specified to decrease class entropy (that is, increase classification accuracy), and (2) the problem solution complexity k_d, which specifies the maximal number of attributes necessary to classify a problem instance accurately. The former is relevant for supply and growth. The latter is relevant for the sustenance of a complete problem solution.

Putting the bounds together, we showed that XCS can *PAC-learn* a restricted class of *k-DNF* problems. However, the reader should keep in mind that XCS is an evolutionary-based, online generalizing RL system and is certainly not designed to learn *k-DNF* problems particularly well. In fact, XCS can learn a much larger range of problems including *k-DNF* problems but also multistep RL problems, real-valued problems, datamining problems, or function approximation problems as validated in subsequent chapters.

Before the validation, though, we need to investigate the last three points of our facetwise LCS theory approach for single-step (classification) problems. Essentially, it needs to be investigated if search via mutation and recombination is effective in XCS and how XCS distinguishes between local and global solution structure. The next chapter consequently considers problems which are hard for XCS's search mechanism since whole BB structures need to be processed to evolve a complete problem solution. We consequently improve XCS's crossover operator using statistical methods to detect and propagate dependency structures (BBs) effectively.

Effective XCS Search: Building Block Processing

The facetwise approach to GA theory stresses effective mixing and decision making among BBs. The last chapter showed that in XCS, BBs are subsets of specified attributes that increase accuracy. The reproductive opportunity bound additionally ensures that BBs are able to grow in the population making time for the identification and reproduction of schema representatives. Until now, we assumed that mutation is sufficient to generate better classifiers as investigated in the time bound. However, the GA literature suggests that effective crossover operators are mandatory to solve boundedly difficulty optimization problems in which small, lower-level BBs may mislead the population to a local optimum.

Thus, this section investigates problems that pose a similar BB-challenge to the XCS system. We create hierarchical classification problems that demand the effective processing of lower level BB structures. In effect, we face the third part of the proposed facetwise problem decomposition for LCS systems, that is, the necessity to enable optimal solution search: (1) Search via mutation needs to be effective; (2) Search via recombination needs to be effective; (3) Local problem solution structure may be different from global structure and thus needs to be taken into account when designing effective recombinatory search operators.

Search via mutation was investigated in several of the previous chapters. We showed its influence on specificity as expressed in the specificity equation (Equation 5.8) as well as its influence in generating schema representatives and finding an optimal problem solution (time extension of schema supply bound, time bound, Chapter 6). We also noted slight disruptive effects affecting the sustenance of problem solutions.

This chapter focuses on recombination as well as differences between local and global problem structure. To investigate the effectiveness of recombination, we identify BB-hard problems in classification problems. XCS is not able to solve these problems due to disruption caused by crossover. Mutation alone may solve the problem, but may take a long time. To solve the problems effectively, a competent crossover operator is necessary that recombines BBs without disrupting them. Experiments with an informed crossover operator confirm this hypothesis but are unsatisfactory since BB structures cannot be assumed to be known beforehand.

Thus, we integrate structure extraction mechanisms previously successfully applied in the GA literature. However, since XCS reproduces in action sets and thus in problem subspaces, the methods need to be modified for the XCS system—respecting the difference in local problem solution structure in comparison to global problem solution structure as suggested in the eighth point of our facetwise LCS theory approach.

In particular, we introduce the formation of *marginal product models* used in the *extended compact GA* (ECGA) (Harik, 1999). The technique is able to identify non-overlapping dependency structures in a problem. Since the marginal product model can only model non-overlapping BBs, we also utilize dependency structures in the form of Bayesian decision trees as used in the Bayesian optimization algorithm (BOA) (Pelikan, Goldberg, & Cantu-Paz, 1999). The Bayesian model can also detect overlapping dependency structures.

We integrate the methods in XCS by extracting a dependency structure from the *global* population and using the gained structural knowledge to generate *local* offspring. The resulting enhanced XCS system is able to solve the identified BB-hard problems.

The remainder of this chapter first derives BB-hard problems for classification. The evaluations show that only an informed crossover operator can solve the problems reliably. Next, we introduce competent crossover operators derived from mechanisms used in ECGA and BOA to solve the BB-challenge without any prior structural information. Summary and conclusions put the results into a broader LCS perspective.

7.1 Building Block Hard Problems

As we have seen in the previous chapter, XCS relies on the supply of minimal order schemata that increase classification accuracy—the BBs in XCS. In the previous chapter, we evaluated the schema bound and reproductive opportunity bound in a problem in which one minimal order schema had to be present. The solution was found once the block that specified all k_m attributes correctly was detected and reproduced.

The question now is if XCS is able to identify and process several of those BBs, represented by schemata of order k_m, effectively. Thus, we first revisit fitness guidance to understand BB processing in XCS even better. Next, we create hierarchical classification problems which consist of several BB structures. In order to solve the problems efficiently, it is necessary to identify, reproduce, *and* recombine the blocks appropriately. Thus, fitness guidance needs to be exploited to successfully grow blocks. Effective recombination operators need to be available as well to successfully combine blocks.

This section first provides further exemplar problems and the consequent fitness guidance in the problem. Next, we introduce hierarchical classification problems showing that a proper BB propagation algorithm is mandatory to solve these types of problems effectively. Section 7.2 introduces explicit BB-identification and propagation mechanisms to XCS.

7.1.1 Fitness Guidance Exhibited

As noted before, the strength of the fitness pressure in XCS depends on the problem at hand. A typical easy problem for the XCS mechanism is the count ones problem (Butz, Goldberg, & Tharakunnel, 2003), in which the majority of ones (or zeros) in the relevant attributes decides the class. The accuracy structure in the count ones problem is very similar to the fitness structure of a one-max problem in the GA literature. Each relevant bit raises accuracy. Thus, each relevant bit is progressively more specialized in the condition parts of the classifiers in XCS.

Table 7.1 shows some exemplar classifier condition parts and the corresponding average reward prediction and reward prediction error estimates for classifiers with action part 1 in the count ones problem. It can be seen that the specialization of progressively more ones or more zeroes decreases error and consequently fitness. Thus, fitness progressively pushes towards the specification of more ones (zeros) in the problem. Butz, Goldberg, and Tharakunnel (2003) showed that uniform crossover can assure and improve successful learning of the count ones problem with many additional irrelevant bits due to its effective uniform recombination.

Table 7.1. Expected reward prediction and reward prediction error estimates for exemplar condition parts in several typically-used Boolean function problems for classifiers with action part $A = 1$.

5-Count-Ones Problem			Hidden 4-Parity Problem			6-Multiplexer Problem		
C	R	ε	C	R	ε	C	R	ε
#####	500.0	500.0	#####	500.0	500.0	######	500.0	500.0
1####	687.5	429.7	1####	500.0	500.0	1#####	500.0	500.0
##1##	687.5	429.7	0####	500.0	500.0	0#####	500.0	500.0
0####	312.5	429.7	11###	500.0	500.0	##1###	625.0	468.8
####0	312.5	429.7	1##1#	500.0	500.0	##0###	375.0	468.8
11###	875.0	218.8	00###	500.0	500.0	##11##	750.0	375.0
##1#1	875.0	218.8	111##	500.0	500.0	##00##	250.0	375.0
00###	125.0	218.8	000##	500.0	500.0	0#1###	750.0	375.0
#0#0#	125.0	218.8	101##	500.0	500.0	0#0###	250.0	375.0
110##	750.0	375.0	1110#	1000.0	0.0	0#11##	1000.0	0.0
111##	1000.0	0.0	0100#	1000.0	0.0	001###	1000.0	0.0
11##1	1000.0	0.0	0000#	0.0	0.0	10##1#	1000.0	0.0
000##	0.0	0.0	1010#	0.0	0.0	000###	0.0	0.0
0##00	0.0	0.0	1111#	0.0	0.0	01#0##	0.0	0.0

In comparison with the count-ones problem, the hidden parity problem (Kovacs, 1999) is harder because the specialization of one attribute of the parity bits does not raise accuracy. Only once all relevant attributes are specialized, accuracy raises, effectively directly solving the problem. Thus, supply of classifiers that specialize all parity bits is necessary, as also shown in the previous chapter. Table 7.1 shows the four hidden parity problem (the fifth bit is irrelevant). Error only drops to zero once all four attributes are correctly specified. In the next section we show that a hierarchical parity, multiplexer problem forces XCS to propagate the lower level parity blocks effectively.

Finally, we again show the widely studied multiplexer problem (Wilson, 1995; Wilson, 1998), in which accuracy somewhat guides towards the correct specializations. Initially, though, only the specialization of the value bits raises accuracy. It is only once some value bits are specified in a classifier condition that specialization of the address bits decreases accuracy further. Table 7.1 clarifies the property in the 6-multiplexer case. When starting with complete generality ($P_\# = 1.0$), relying on mutation for the first specializations, specificity initially raises more in the value attributes of the classifiers. Only later does specificity in the address attributes take over.

7.1.2 Building Blocks in Classification Problems

The above problems consist of BB structure that either consist of only one BB, as in the hidden parity problem, or many single-attribute BBs as in the count ones problem. The multiplexer problem is somewhat a hybrid since initial fitness guidance leads to the less important value bits and only later, fitness guides towards the specialization of the address bits. Thus, BB processing is somewhat easy in the count ones problem in that only one specialization needs to be identified at a time. This can be accomplished by mutation. More challenging is the hidden parity problem in which classifiers that specialize all relevant bits need to be available from the beginning.

What if we combine the output of multiple hidden-parity problems to a higher-level problem input? The hierarchical dependency between the subproblems would require that the hidden parity blocks need to be identified and then recombined effectively. This described hierarchical problem structure consequently requires effective BB processing.

We construct such a problem structure using a two-level hierarchy. On the lower level, small Boolean functions are evaluated which provide the input to the higher level. Thus, the function evaluation is pursued in two stages. The evaluation of the functions on the lower level serve as input to the higher level. For example, we mainly use a parity, multiplexer combination in which the small lower-level blocks are evaluated by the parity function. The results are then fed into the higher-level multiplexer function deriving the overall class of the problem instance. Further information and visualizations on the hierarchical problem class are provided in Appendix C.

Note that we are not interested in creating a problem to force BB processing for its own sake. In fact, many indications in nature and engineering suggest that typical natural problems are structured in a hierarchical, decomposable structure (Simon, 1969; Gibson, 1979; Goldberg, 2002). Thus, we believe that the introduced hierarchical problem is an important problem to solve with a general machine learning system.

How can XCS solve this problem? Clearly, the lower level parity blocks need to be identified first to enable the discovery of the higher level function. Table 7.2 shows exemplar conditions with corresponding average reward predictions and prediction errors for the hierarchical 3-parity, 6-multiplexer problem. In contrast to the plain multiplexer problem or count ones problem, in these hierarchical problems the lower-level BBs (for example, parity blocks) need to be identified and then processed effectively. The next section shows that it is only if the detected blocks are not disrupted that XCS is able to solve the problem. Additionally, it is only if the BBs are recombined effectively that XCS can solve the problem efficiently.

Table 7.2. Expected reward prediction and reward prediction error measures for exemplar condition parts in the hierarchical 3-parity, 6-multiplexer problem for classifiers with action part $A = 1$. For readability reasons, the lower level 3-parities are tightly coded and separated by spaces.

C	R	ε
### ### ### ### ### ###	500.0	500.0
111 ### ### ### ### ###	500.0	500.0
#1# ### #11 #1# #11 ###	500.0	500.0
### ### 111 ### ### ###	625.0	468.8
### #1# ### 100 ##1 ###	625.0	468.8
### 0## ### ### 000 ###	375.0	468.8
### 111 ### 010 ### ###	750.0	375.0
##1 111 ##0 100 #0# ###	750.0	375.0
101 ### 111 ### ### ###	750.0	375.0
### 000 ### ### 000 ###	250.0	375.0
101 111 ### 100 ### ###	1000.0	0.0
101 000 111 ### ### ###	1000.0	0.0

Note that we focus in the remainder on XCS's performance in the parity, multiplexer and parity, count-ones combination. Nonetheless, any other type of Boolean function combination in the proposed hierarchical manner is possible. Additionally, it is not necessary that all BBs on the lower level are evaluated by the same Boolean function, nor do they need to be of equal length. Certainly, though, all these potential manipulations may lead to different challenges with respect to the facetwise theory for LCSs.

7.1.3 The Need for Effective BB Processing

We tested XCS on the proposed hierarchical problem combining parity and multiplexer problem as well as parity and count ones problem. Results confirm that the parity, multiplexer combination is particularly challenging.

Hierarchical Three-Parity, Six-Multiplexer Problem

Performance of XCS in the hierarchical 3-parity, 6-multiplexer problem is shown in Figure 7.1.[1] It can be seen that XCS is not able to solve the problem if uniform crossover is applied. Due to the disruptive effects of uniform crossover—as already suggested in Holland's original schema theory (Holland, 1975)—XCS is not able to process the lower level BBs, but rather disrupts them.

In addition to the usual crossover operators of uniform, one-point, and two-point crossover, we applied an informed crossover operator to investigate the potential of more competent recombination operators. The informed crossover operator is informed about the BB structure in the problem applying a BB-wise uniform crossover operator. BB-wise uniform crossover exchanges only complete BBs uniformly randomly similar to uniform crossover, which exchanges attributes uniformly randomly.

XCS with BB-wise crossover solves the problem effectively and nearly independently of the mutation type used. Thus, a mechanism in XCS that is able to identify the lower-level BB structure is necessary. Once identification is successful, effective BB processing and recombination can be applied. Uniform crossover strongly disrupts BB propagation, preventing learning (Figure 7.1a,c). The continuously high population size indicates that uniform crossover causes a high diversity in the population. However, BB-disruption prevents the growth of higher-order BBs (Figure 7.1b,d). Mutation alone is able to solve the problem but learning takes about three times as long as in the case of BB-wise uniform crossover operator (Figure 7.1a,c). The population sizes indicate that diversity stays much lower resulting in a lower macro-population size (Figure 7.1a,c). If the BBs are tightly coded, one-point and two-point crossover are able to recombine BBs effectively (Figure 7.1a). However, if the attributes are randomly distributed, the potential recombinatory benefit of one-point or two-point crossover is overshadowed by their disruptive effect delaying learning and population convergence (Figure 7.1c,d). Thus, one-point and two-point crossover show beneficial effects in the case of tightly-coded BBs, but disruptive effects in the loosely-coded case. Since the

---

[1] If not stated differently, all results in this chapter are averaged over ten experiments. Performance is assessed by test trials in which no learning takes place and the better classification is chosen. During learning, classifications are chosen at random. If not stated differently, parameters were set as follows: $N = 20000$, $\beta = 0.2$, $\alpha = 1$, $\varepsilon_0 = 10$, $\nu = 5$, $\theta_{GA} = 25$, $\chi = 1.0$, $\mu = 0.01$, $\gamma = 0.9$, $\theta_{del} = 20$, $\delta = 0.1$, $\theta_{sub} = 20$, and $P_\# = 0.6$.

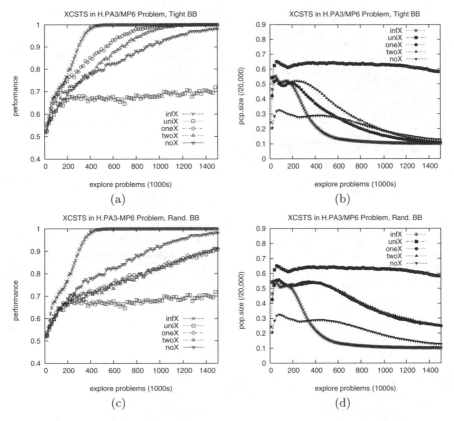

Fig. 7.1. Performance (a,c) and population sizes (b,d) of XCS ($N = 20k$) in the hierarchical 3-parity, 6-multiplexer problem (infX = informed (i.e. BB-wise uniform) crossover, uniX = uniform crossover, oneX = one-point crossover, twoX = two-point crossover, noX = mutation only). Efficient BB recombination strongly improves XCS's performance. One-point and two-point crossover are only beneficial if the BBs are tightly coded. Although mutation alone is able to solve the problem, the time until the solution is found is much larger.

dependency structures cannot be expected to be tightly coded in general, competent crossover operators are mandatory.

Although mutation can be tuned to solve the hierarchical 3-parity, 6-multiplexer problem nearly as well as the informed crossover operator does (Figure 7.2b), the behavior is unsatisfactory: larger problems or the same problem with additional irrelevant attributes would make it impossible to set the mutation rate high enough due to the reproductive opportunity bound introduced in the last chapter. However, a small mutation rate strongly delays learning if only mutation is applied. The BB-wise uniform crossover operator stays nearly independent from the mutation operator (Figure 7.2a,b). It only relies on the supply of lower level BBs, which is usually ensured by the initial

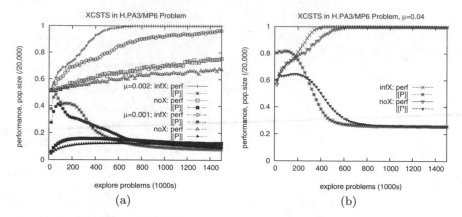

Fig. 7.2. A low mutation rate strongly delays learning if effective recombination is not applied. With a mutation rate of $\mu = 0.001$, however, certain specialized attributes might get lost so that performance is delayed even with effective recombination (a). Higher mutation rates alleviate the problem (b) but may not be applicable in problems with more attributes.

sufficiently large specificity. Thus, a competent crossover operator that detects BBs on the fly is highly desirable.

Hierarchical Parity, Count-Ones Problem

Figure 7.3 confirms similar results in the hierarchical 3-parity, 5-count ones problem. In the runs, population size is set to $N = 20,000$. Note that the problem has as many niches as the 3-parity, 6-multiplexer problem. However, the smaller population size as well as the overlapping niches in the problem make it very hard for XCS to solve the problem completely optimally. Nonetheless, effective recombination significantly improves performance. As before, one-point and two-point crossover are only effective if the blocks are tightly coded. Otherwise, the operators are nearly as disruptive as uniform crossover. Mutation alone slowly improves performance but takes a very long time to evolve an accurate solution. The performance of BB-wise crossover is not reached by any of the other settings.

7.2 Building Block Identification and Processing

Facing the BB-challenge within XCS it is necessary to develop a mechanism that learns effective recombination online. Most appropriate for this seems to be an estimation of distribution algorithm (EDA) approach, modeling dependency structures and recombining them appropriately (Pelikan, Goldberg, & Lobo, 2002; Larrañaga, 2002).

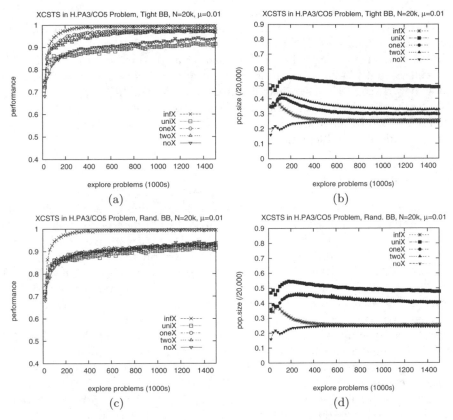

Fig. 7.3. Performance (a,c) and population sizes (b,d) of XCS ($N = 20k$) in the hierarchical 3-parity, 5-count ones problem. Again, efficient BB recombination strongly improves XCS's performance. One-point and two-point crossover are only beneficial if the BBs are tightly coded. Mutation alone gradually improves performance but is much less effective that BB-wise crossover.

However, the evolutionary component in XCS differs from the usual GA application in several respects. Due to XCS's niche reproduction in action sets and since action sets are generally rather small compared to the whole population, structure extraction is hard to apply successfully in an action set alone. On the other hand, extracting global structure results in global offspring generation, which may not reflect the local problem structure appropriately. Thus, the inclusion of an EDA mechanism in XCS is not straight-forward.

This section integrates the BB-identification mechanism in the ECGA (Harik, 1999) to identify and process lower-level dependency structures. Alternatively, we also show how to integrate the more powerful Bayesian learning mechanism used in BOA (Pelikan, Goldberg, & Cantu-Paz, 1999). We show that both mechanisms are suitable to learn the *global* lower-level problem structure and can be used to generate or improve *local* classifier offspring. The generation and improvement of the local offspring depends on the current action set, just as the original XCS crossover operation does.

We first give an overview of the learning algorithms used in the ECGA as well as in BOA to learn the respective dependency structures. Next, we show how these mechanisms may be integrated into the XCS classifier system. We learn the dependency structures from the filtered and converted XCS population and then use the learned structures to sample and/or optimize offspring in local problem niches.

7.2.1 Structure Identification Mechanisms

Our investigations show that at least two structure identification mechanisms are suitable for competent BB processing in XCS: (1) *marginal product models* used for example in the ECGA mechanism (Harik, 1999), and (2) Bayesian decision tree structures used in BOA (Pelikan, Goldberg, & Cantu-Paz, 1999; Pelikan, 2002). The former is easier to understand and to apply but is limited to the identification of non-overlapping BBs only. The latter is more complicated but is able to model overlapping dependency structures as well.

Any structure extraction mechanism, however, faces the problem of accuracy vs. generality. That is, the generated model is intended to identify relevant dependencies but ignore spurious, irrelevant dependencies. Hereby, we rely on *Occam's razor* in that we want to find the model that codes the data structure most compactly. The usual approach to balance the two optimization factors is to apply the minimum description length principle (MDL) (Mitchell, 1997). Essentially, the MDL principle weighs accuracy with model complexity by combining the cost of describing the derived model with the resulting cost of encoding the modeled data using the model. Using information theoretic principles, the two influences can be appropriately balanced using entropy as the basic measure.

7.2.2 BB-Identification in the ECGA

As mentioned above, the ECGA mechanism learns a non-overlapping BB-structure, termed a marginal product model. ECGA considers the best individuals in its current population (selected by any suitable selection mechanism, e.g. tournament selection) and builds the model from these individuals, that is, the data. For example, consider the simple population shown in Table 7.3. ECGA finds dependency structures in terms of feature subsets (that is, the BBs). A block is essentially formed if the representation as a block, albeit more complex to express as a model, results in a sufficient reduction in the resulting data description complexity when using the model.

ECGA expresses these two complexity measures in terms of model complexity MC, which favors more compact models, and the resulting compressed population complexity CPC, which favors a more compact (accurate) population representation with respect to the used model. The MDL measure is simply the sum of MC and CPC. The two complexity measures are determined by

$$MC = \log N \sum_I 2^{S[I]} - 1, \qquad (7.1)$$

$$CPC = N \sum_I E(M_I), \qquad (7.2)$$

where N specifies the population size, I a dependency subset, $S[I]$ the number of attributes in subset I, M_I the probability distribution of all possible values in subset I, and $E(M_I)$ the entropy of a probability distribution. The measure MC exploits the fact that $\log N 2^{S[I]}$ bits are necessary to describe the probability distributions over each subset $S[I]$ ($2^{S[I]}$) probability entries. The measure CPC then determines the complexity of coding all N individuals with respect to the subsets, which is determined by the sum of the entropies over all subsets. Table 7.3 shows several potential model structures and the resulting MC and CPC measures. It can be seen how the MDL principle balances the model complexity with the resulting population description complexity.

Table 7.3. The illustrated example shows how the marginal product model learning mechanism detects structural properties in a population. While MC measures the model complexity, CPC measures the compressed population complexity potentially gained due to a more complex model representation.

problem instances		marginal product model	MC	CPC	MC+CPC
11000	11111	[1][2][3][4][5]	15	36.98	51.98
11001	11110	[1 2] [3] [4] [5]	18	30.49	48.49
11000	11110	[1 2] [3 4] [5]	21	22.49	43.49
00111	00001	[1 2 3 4] [5]	48	22.49	70.49
		[1 2 3] [4 5]	30	30.49	60.49

The ECGA mechanism learns the marginal product model, greedily minimizing the sum of MC and CPC. That is, if the scaled entropy decrease and thus the decrease of CPC due to a merge of two sets is larger than the consequent MC increase, the merge is performed. Subsets are greedily merged until no more merge is able to decrease the MDL measure. In the ECGA, the model is learned every GA iteration. The offspring population is generated out of the derived dependency structure probabilistically sampling from the dependency structure. That is, each BB is considered independently when generating an offspring individual choosing the corresponding code probabilistically with respect to the determined probability distribution. The MDL mechanisms used to grow the dependency structure in XCS similar to the ECGA are taken from the available ECGA implementation (Lobo & Harik, 1999).

The ECGA mechanism has shown to be able to effectively solve previously BB-hard problems, such as the typically used deceptive trap problems (Harik, 1999; Sastry & Goldberg, 2000). Due to its rather straight-forward approach and the various successful applications, it appears a valuable candidate for integration into XCS.

7.2.3 BB-Identification in BOA

BOA uses the more powerful representation of Bayesian networks in order to represent BB-structures. The overall learning mechanism is similar to the one applied in the ECGA, learning a Bayesian network from a selected subset of individuals and sampling from the Bayesian network. Due to the potentially much more complex Bayesian network structure, the generation and sampling mechanisms are not as straight-forward as in the ECGA.

Bayesian networks (BNs) (Howard & Matheson, 1981; Pearl, 1988; Buntine, 1991; Mitchell, 1997) combine statistics with graph theory generating a modular graphical model of the analyzed data. BNs can be used to estimate probability distributions as well as to do inference. A Bayesian network is defined by its structure and its (conditional) probabilities. The structure is usually encoded by a directed acyclic graph with the nodes corresponding to the features and the edges corresponding to conditional dependencies. The parameters are represented by a set of conditional probability tables (CPTs) specifying a conditional probability for each variable given any instance of the variables that the variable depends on.

The BN as a whole encodes a joint probability distribution given by

$$p(x) = \prod_{i=1}^{n} p(x_i | \Pi_i), \qquad (7.3)$$

where $X = (X_0, \ldots, x_{n-1})$ is a vector of all the variables in the problem; Π_i is the set of parents of x_i (the set of nodes from which there exists an edge to x_i); and $p(x_i | \Pi_i)$ is the conditional probability of x_i given its parents Π_i. A CPT then codes the probability of the values of x_i given the parental values. A directed edge relates the variables so that in the encoded distribution, a variable corresponding to a terminal node is conditioned on the parental variables. More incoming edges into a node result in a conditional probability of the variable with a condition containing all its parents.

As the ECGA structure assumes the independence of the blocks, also a Bayesian network encodes a set of (implicit) independence assumptions. Variables are assumed to be independent of each other given the values of the variables of all of their parents and none of their common descendants. The exact independence assumptions resulting from the BN structure can be found in the literature (Mitchell, 1997).

Conditional probability tables (CPTs) store the conditional probabilities $p(x_i | \Pi_i)$ for each variable x_i. The number of conditional probabilities for a variable that is conditioned on k parents grows exponentially with k. For binary variables, for instance, the number of conditional probabilities is 2^k, because there are 2^k instances of k parents and it is sufficient to store the probability of the variable being 1 for each such instance. Figure 7.4 shows an example CPT for $p(x_1|x_2, x_3, x_4)$.

A greedy algorithm is usually used to learn a BN. The greedy algorithm starts with an empty BN. Each iteration, an edge is added to the network that improves quality of the network maximally. Network quality can be measured by any popular scoring metric for Bayesian networks, such as the Bayesian Dirichlet metric with likelihood equivalence (BDe) (Cooper & Herskovits, 1992; Heckerman, Geiger, & Chickering, 1994) or the Bayesian information criterion (BIC) (Schwarz, 1978). Learning terminates when no more improvement is possible.

The sampling of a Bayesian network can be done using probabilistic logic sampling (PLS) (Henrion, 1988). In PLS the variables are ordered topologically so that it is assured that every variable is preceded by all parental variables it depends on. Variable values are then generated iteratively according to the topological ordering. As a result, once the value of a variable x_i is to be generated, the values of its parents Π_i are assured to have been generated already. Thus, the probabilities of different values of x_i can be directly extracted from the CPT for x_i using the known values of Π_i.

Despite the encoded independence assumptions in a Bayesian network, identified dependencies may also contain regularities. Furthermore, the exponential growth of full CPTs with respect to the number of parents often obstructs the creation of models that are both accurate and efficient. Thus, often local structures are used in Bayesian networks to represent local conditional probabilities more efficiently than traditional full BNs (Chickering, Heckerman, & Meek, 1997; Friedman & Goldszmidt, 1999).

Pelikan (2002) uses decision trees to store the conditional probabilities of each variable in a separate tree. Each internal (non-leaf) node in the decision tree for $p(x_i|\Pi_i)$ has a variable from Π_i associated with it and the edges connecting the node to its children stand for different values of the variable. For binary variables, there are two edges coming out of each internal node; one edge corresponds to 0 and the other edge corresponds to 1. For more than two values, either one edge can be used for each value, or the values may be classified into several categories and each category creates an edge.

Each path in the decision tree for $p(x_i|\Pi_i)$ that starts in the root of the tree and ends in a leaf encodes a set of constraints on the values of variables in Π_i. Each leaf stores the value of a conditional probability of $x_i = 1$ given the condition specified by the path from the root of the tree to the leaf. A decision tree can encode the full conditional probability table for a variable with k parents if it splits to 2^k leaves, each corresponding to a unique condition. However, a decision tree enables the more efficient and flexible representation of local conditional distributions. See Figure 7.4b for an example decision tree modeling the conditional probability table presented earlier.

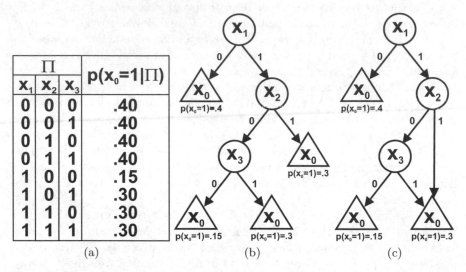

Π			$p(x_0=1|\Pi)$
x_1	x_2	x_3	
0	0	0	.40
0	0	1	.40
0	1	0	.40
0	1	1	.40
1	0	0	.15
1	0	1	.30
1	1	0	.30
1	1	1	.30

(a)   (b)   (c)

Fig. 7.4. A conditional probability distribution representation for $p(x_0|x_1, x_2, x_3)$ using a full-blown conditional probability table (a), as well as a decision tree (b) and a decision graph (c).

Pelikan (2002) uses also the (acyclic) decision graph feature allowing more edges to terminate in a single node, enabling the sharing of children by several internal nodes. This makes the representation even more flexible and allows even more compact dependency structure representations. Figure 7.4c shows an exemplar decision graph.

To learn Bayesian networks with decision trees, a decision tree for each variable x_i is initialized to an empty tree with a univariate probability of $x_i = 1$. In each iteration, each leaf of each decision tree is split (as long as a topological ordering remains possible) determining the quality change of the current network measured by the applied metric. The best split is performed. Learning stops when no potential split is able to improve the current network.

To estimate model quality, a combination of the BDe (Cooper & Herskovits, 1992; Heckerman, Geiger, & Chickering, 1994) and BIC (Schwarz, 1978) metrics is used, where the BDe score is penalized with the number of bits required to encode parameters (Pelikan, 2002). For decision graphs, a merge operation is introduced to allow merging two leaves in any (single) decision graph. The Bayesian decision graph mechanism applied to XCS is taken from Pelikan's BOA implementation available on the net (Pelikan, 2001)

Similar to the ECGA approach, Bayesian networks model dependencies and independencies where a dependency is defined as a *non-linearity*. This non-linearity is identified by an entropy decrease as measured in the applied metric. Applying the introduced decision graph structure focuses the modeled dependencies in one decision graph on one feature modeling local, lower-level dependency structures, that is, the expected BBs in the problem.

7.2.4 Learning Dependency Structures in XCS

Similar to the BB-identification mechanisms in ECGA and BOA, which search BBs in the current best subset of individuals, it is possible to learn dependency structures from the current population in XCS. However, two aspects need to be considered. (1) Selection from the global population is not straight-forward. (2) Classifiers need to be suitably transferred into binary.

As in ECGA and BOA, the dependency structure needs to be built from the current best individuals in the population of XCS. Despite the fitness sharing in XCS, relative accuracy may not be the appropriate measure since different problem niches may currently exhibit different learning stages so that the fitness may be misleading, potentially favoring already converged subspaces. However, it is possible to require a certain classifier confidence for selection, similar to the thresholded application of subsumption. We use a filtering mechanism that extracts the most accurate classifiers out of the current population. The mechanism extracts those classifiers that have a minimum experience θ_{be}, a minimum numerosity θ_{bn}, and a maximum error θ_{be}. The parameters were set to $\theta_{be} = 20$, $\theta_{bn} = 1$, $\theta_{b\varepsilon} = 400$ throughout the subsequent experiments filtering out the young and high-error classifiers. Since predictions below the average reward of 500 can be considered as predictions of the opposite class with higher reward, we switch the class of those classifiers that predict a reward of less than 500. Note that this method can only be applied in classification problem in which only two types of reward (e.g. 1000/0) are possible.

Given a filtered population, how to translate the classifier population into a suitable representation to build the model needs to be clarified. Don't care symbols may be simply coded by a third symbol in a ternary alphabet. However, don't care symbols do have a special meaning in that they match zero or one. Thus, to simplify model-building, we decided to code each condition attribute by two bits: The first bit encodes if the condition attribute is general (that is, don't care) or specific. The second bit encodes the value of the attribute. If the attribute is a don't care symbol, we choose zero or one uniformly randomly for the second bit. Finally, the classification part may yet play a special role and future work may build models for each classification separately. For now, we simply code the classification part as another bit. Table 7.4 shows a set of classifiers and the corresponding encoding that is used to learn the Bayesian network with decision graphs.

With a binary coded set of individuals at hand, we are able to learn the BB structure via the MDL-metric of the ECGA or the Bayesian decision graph structure via the Bayesian-network learning algorithm. Note that since XCS applies a steady-state niche GA, the dependency structure does not need to be rebuilt every time step. We rebuild the network after a fixed number of

Table 7.4. Sample classifiers (from the multiplexer problem) and their corresponding binary encoding for the structure learning mechanism. Spaces are added for clarity. If an attribute is a don't care symbol, the second bit in the corresponding binary code is chosen randomly. The class bit is flipped, if the reward prediction is below 500.

C	A	R	ε	binary encoding
##11##	1	750	375	10 11 01 01 11 10 1
##00##	1	250	375	11 11 00 00 10 11 0
0#1###	0	250	375	00 01 11 11 11 10 1
0#0###	0	750	375	00 00 10 11 10 10 0
0#11##	1	1000	0	00 11 01 01 11 10 1
0#11##	0	1000	0	00 11 01 01 11 11 0
001###	1	1000	0	00 00 01 10 10 10 1
10##0#	0	1000	0	01 00 11 11 00 10 0
000###	1	0	0	00 00 00 11 10 10 0
01#1##	0	0	0	00 01 11 01 11 11 1

time steps θ_{bs}, usually set to $10,000$ in our experiments. The threshold is only slightly problem dependent and does not appear to have a strong impact on performance. In general, the lower the threshold, the more often the model is rebuilt, potentially adjusting the model to newly detected dependencies faster but also potentially wasting computational resources for rebuilding the same dependency structure.

7.2.5 Sampling from the Learned Dependency Structures

As shown in Section 7.1, the recombination of the parents using common crossover operators may lead to disruptive effects potentially destroying important BB-structure. Once the dependency model is learned, XCS may use the model to recombine or directly generate offspring classifiers more effectively. As long as the learned model reflects important dependency structures, it can be expected that the resulting recombination is less disruptive and much more directed towards generating offspring that combines already successfully learned substructures effectively searching in the neighborhoods defined by the substructures.

As investigated in detail in the previous chapters, XCS generates offspring from parental classifiers selected from the current action set. This means that XCS reproduces classifiers that encode solutions with respect to the current problem instance. When using the globally learned dependency structures to generate offspring, we consequently need to adjust the structures to be able to generate local offspring. We investigate the following two options: (1) sampling classifiers using the model updated to the local probability distribution; and (2) probabilistically improving the selected parental offspring classifier using the model with global or local probabilities.

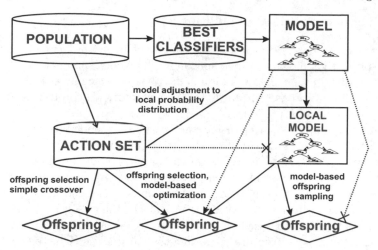

Fig. 7.5. A probabilistic model of the problem structure can be built and used for offspring generation in many ways. Since action sets are usually too small to gather reliable statistics, a selected classifier subset from the global population should reflect problem structure most effectively. Once the model is formed, offspring may either be generated optimizing a parental classifier or sampling directly from the model. Sampling from the global model is inappropriate since the global distribution does not reflect the local solution structure. Setting the model distribution probabilities to the local probability distribution results in a model that encodes global dependency structures with respect to the local probability distribution. Thus, optimizing offspring by the means of the local model can be expected to be most cautious and most robust.

Figure 7.5 shows the different potential methods for offspring generation by the means of a dependency model structure. Since XCS generates offspring in local niches, reproducing classifiers simply sampling from the global model is impossible since the classifier cannot be expected to reflect the solution structure in the current niche. Similarly, optimizing classifier structure by the means of the global model with global probabilities is expected to be disruptive as well since the optimization biased on the global probability structure again generates a classifier that reflects the global probability distribution. Even if the global model represents the dependencies in the population optimally, it may not be used to directly sample local offspring since it can only be expected to code lower-level BB information. Higher-level BB dependencies depend on the problem niche under investigation. Thus, it appears ineffective and very hard to grasp these higher-level dependencies in the global model.

Both offspring generation methods are introduced next. The latter is used only in conjunction with the learned Bayesian networks.

Sampling using Local Probabilities

As shown in Chapter 5, reproducing classifiers in action sets yields classifier offspring that has an average specificity that corresponds to the average specificity distribution in the action sets. Fitness may increase the average specificity due its pressure towards higher accuracy, which often leads to an implicit specialization pressure.

Thus, to sample offspring using the learned dependency structure, the model probabilities need to reflect the local specificity distribution. Consequently, we update the probabilities in the applied model with respect to the best classifiers in the current action set. To achieve this, we select a subset of classifiers from the action set using the tournament selection mechanism. The selected subset (which may contain identical classifiers several times selecting with replacement) is used to update the (conditional) probabilities. The updated dependency structure consequently reflects the detected global dependency structure but mimics the local probability distribution. The globally detected dependencies are thus combined with the local probabilities resulting in an offspring sampling mechanism that combines global with local problem knowledge.

The sampling is then achieved using the sampling mechanisms in the dependency structures explained above. Effectively, we use the globally detected dependency structures to sample local offspring. Hereby, the globally extracted structure biases the recombination. The current local probability distribution is used to sample offspring locally using the global structure. As a result, we recombine the locally important BBs effectively as long as the global dependency structure applies in the local problem niche.

Structure Optimization

In the former case, we combined global model information about dependencies with the local probability distribution. Vice versa, it is also possible to use the global dependency structure and probability distribution to optimize the local probability structure. Hereby we have to be cautious to not overwrite the local information completely by the global information.

Essentially we apply a Markov Chain Monte Carlo (MCMC) approach (Neal, 1993) first introduced in the statistical physics literature in the 1950s in the so-called *Metropolis Algorithm*. An MCMC essentially iteratively and probabilistically changes a current probability distribution to an equilibrium distribution. MCMC iteratively evaluates potential changes of single attributes and decides probabilistically if the change should be made. Which probabilities are chosen to confirm an update is application dependent. Our method uses the likelihood of the change with respect to the global model.

Particularly, XCS chooses an offspring via tournament selection in the current action set. Instead of simple crossover, XCS then applies the MCMC mechanism to probabilistically optimize the classifier structure. Bits of the binary code of a classifier are chosen at random determining the likelihood of the structure before and after the change. To avoid zero likelihoods, all conditional probabilities are linearly normalized to values ranging from 0.05 to 0.95. Normalizing the likelihood before and after the change to one, the change is then committed with the probability of the normalized likelihood.

In effect, the MCMC pushes the selected local classifier towards the global probability distribution. The aim is to combine local *and* global structural information in the offspring classifier. Too many update iterations can be expected to not be useful since the resulting classifier will reflect the global model structure. On the other hand, too few updates will have no effect at all. The subsequent experimental evaluations confirm these expectations.

To avoid using the global probability distribution, it is also possible to combine the two mechanisms above, adjusting the dependency structure to reflect the local probability distribution using the consequent structure to probabilistically optimize a selected local offspring classifier. This has the advantage of avoiding the problem of over-optimization towards the global structure. Additionally, the freedom of sampling local offspring is further constrained since a parental offspring classifier is optimized, constraining the search to an actual parental classifier. In effect, the combination might be the most cautious but also the most robust offspring optimization mechanism overall.

7.2.6 Experimental Evaluation

We evaluate XCS's performance on the above introduced hierarchical problems evaluating and comparing both offspring generation methods applying several different settings. XCS is set to learn a Bayesian network every 10,000 learning steps ($\theta_{bs} = 10,000$). The population XCS learns from is the filtered population as explained above. If the filtered population is empty, no model is learned. As long as no model is learned, XCS applies uniform crossover instead of the model-based crossover. Mutation is applied to the offspring classifiers generated by the model, as before when simple crossover was applied. The results are averaged over ten experiments. Other parameters are set as above except for the population size which is set to $N = 20,000$ in all runs as well as mutation which is set to $\mu = 0.01$ in the runs with normal crossover operators, to $\mu = 0.001$ in the XCS/BOA combination, and to either value as indicated in the subsequent figures in the XCS/ECGA combination. This population size seems large but the investigated problems are huge as well. For example, the hierarchical 3-parity, 6-multiplexer problem requires a final solution of 2^{10} classifiers so that the only twenty times larger population size appears reasonable.

Hierarchical Parity, Multiplexer Problems

In Section 7.1 we saw that the evolution of a successful solution in the 3-parity 6-multiplexer problem strongly depends on the choice of mutation rate and crossover type. If a small mutation is chosen, effective BB recombination is mandatory and was achieved by an informed BB-wise crossover operator. If mutation is large, the problem was solvable but took longer than with the application of the informed crossover operator. However, we know that a large mutation rate is not an option in larger problems in which a smaller mutation rate is necessary to satisfy the covering challenge as well as the reproductive opportunity bound (see Chapter 6). Thus, to solve the addressed problem, mutation needs to be set low and crossover needs to be effective.

Figure 7.6 shows XCS's performance in the hierarchical 3-parity, 6-multiplexer problem. In the ECGA comparison (Figure 7.6 a,b), we see that while BB-wise crossover learns the model slightly faster, ECGA reaches similar performance. The different settings refer to the number of selected classifiers used to adjust the model to the local probability distribution. Higher mutation rates are actually somewhat disruptive as also indicated by the resulting higher population sizes.

An additional specializing effect is observable, which is partially a result of the binary recoding of the population for the model building and model-based offspring generation. Due to the random choice of the second bit of a don't care attribute, actual offspring may be generated that does not match the current action set. For example, consider the two simple classifiers 1#→1 ##→1. The resulting binary codes may be 0111 and 1010. Thus, dependent on the model dependencies, offspring may be generated with the code 0010, which translates into 0#→1. Although the average specificity is maintained, the offspring may not match the current action set, consequently increasing diversity in the population. This additional diversity may be slightly disruptive as observed in the ECGA graphs. Note that we also ran experiments applying uniform crossover on the transferred binary code. The result was that the population was filled up with apparently meaningless classifiers. No learning was observable in this case.

The application of the Bayesian model results in a similarly successful solution of the 3-parity, 6-multiplexer problem. The probabilistic optimization method even reaches higher performance slightly faster than the informed BB-wise crossover application (curve BOA: 0/18). However, if too many optimization steps are applied (0/90), the mechanism over-optimizes the offspring towards the global probability distribution and consequently over-specializes the population with respect to the current global model. This problem does not occur when offspring is sampled using the local probability distribution or if selected offspring is optimized using the local probability distribution. All settings exhibit similar performance nearly as good as the informed BB-wise crossover technique. In contrast to the ECGA combination, the BOA combination does not suffer from any over-specializations and all runs reach 100% performance reliably.

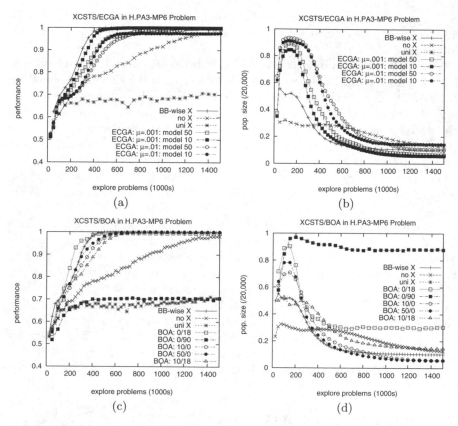

Fig. 7.6. When applying ECGA or BOA to the hierarchical 3-parity, 6-multiplexer problem, recombination becomes effective and XCS is able to effectively learn a complete solution comparatively fast to the runs with BB-wise crossover. The application of the ECGA-based model learning mechanism (a,b) shows competent performance. The 50/10 variation refers to the number of selected classifiers used to set the probabilities to the local probability distribution. In the BOA-based model learning (c,d), the first number again refers to the number of selected classifiers used to set the probabilities to the local distribution (0 indicates that the global probability distribution is used). The second number refers to the probabilistic optimization steps applied to a selected parental classifier (0 indicates that offspring was sampled directly from the model).

Hierarchical Parity, Count Ones Problem

Also the investigated hierarchical 3-parity, 5-count ones problem requires a final optimal solution size of 2^{10}. However, in this case the final population size is overlapping in that three out of five parity blocks need to be specified correctly to predict class zero or one accurately. XCS with ECGA model does not show any problems in solving the problem (Figure 7.7a,b). All runs converge to the near-optimal solution nearly as fast as the informed BB-wise crossover runs. Even with a lower mutation rate, performance is hardly influenced.

Similarly, the XCS runs with Bayesian network successfully learn the problem (Figure 7.7c,d). BOA learns slightly slower than the ECGA combination early in the run but then reaches a slightly higher performance level. Apparently, BOA initially models spurious dependencies that may slow down the overall learning process. Since in this problem the propagation of all five BBs independently is nearly most effective, the Bayesian learning algorithm appears to over-model and thus delay learning early on. In the end of a run BOA decreases disruptive effects. Performance of both methods clearly outperforms the runs without crossover application as well as the runs with uniform crossover application.

In sum, the results confirm that XCS can be successfully combined with a number of structural learners to improve offspring generation. The implemented XCS/ECGA and XCS/BOA combinations showed to be able to achieve performance similar to the performance with BB-wise uniform crossover, which relies on explicit problem knowledge. XCS/ECGA as well as XCS/BOA do not require any global problem knowledge and thus allow XCS to flexibly adjust its recombination operators dependent on the encountered problem. The next chapter provides further evidence for the generality of the model-building approach in XCS applying the techniques to several other typical Boolean function problems.

7.3 Summary and Conclusions

This chapter considered the last three aspects of the facetwise LCS theory approach in the XCS learning system for single-step RL problems. We showed that—as in GAs—the problem may require effective BB processing in LCSs to ensure reliable learning of a problem solution. Additionally, we highlighted the difference between LCSs and GAs in that problem structure may differ dependent on which problem subspace is currently under investigation. In essence, different attributes may be relevant in different problem structures so that different recombinatory mechanisms need to be applied dependent on the current problem subspace.

To investigate the recombinatory capabilities of the XCS system, we introduced hierarchical binary classification problems combining parity blocks on the lower level with the multiplexer or count ones function on the higher level. XCS with simple crossover is not able to solve the resulting problems reliably. We observed the expected disruptive effects of simple recombination when applying uniform crossover as well as when applying one-point or two-point crossover with loosely coded blocks (randomly distributed over the population). Mutation alone is able to solve the parity, multiplexer problem

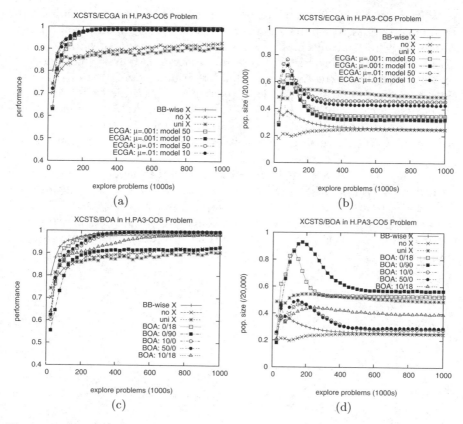

Fig. 7.7. Also the hierarchical 3-parity, 5-count ones problem is effectively solvable with either model-based offspring generation method. The ECGA combination appears slightly more robust in this case, indicating that the Bayesian-net might model unnecessary, spurious dependencies that delay convergence. Note also how over-specialization in the 0/90 setting again disrupts learning.

combination—albeit severely delayed in time—but it is not able to solve the parity, count ones problem combination satisfactorily in the available time.

The integration of the model-building and offspring generation mechanisms from the extended compact GA (ECGA) or the Bayesian model building algorithm (BOA) show that competent crossover operators can be integrated in the XCS learning structure. Since XCS applies a steady-state niche GA, the probabilistic model is not built every time step but it is built from the global model at predefined points in time. The model is then modified to reflect the local probability distribution in an action set at each time step to generate offspring respecting the local probability distribution but biasing recombination on the globally detected dependency structure. XCS combined with either model learner evolved complete solutions to the hierarchical problems.

With respect to the final points of our facetwise theory approach for LCSs we showed that XCS respects the difference between global and local problem structure by reproducing in action sets. Recombination can be made more efficient by using global BB structure information. However, since the local problem structure needs to be respected but usually cannot be coded in the global dependency structure, offspring recombination needs to be adapted to the current local problem niche. Since knowledge about the current niche is represented in the local classifier population, that is, the current action set, model-based offspring generation needs to be biased on the classifier distribution in the current action set.

In conclusion, XCS with the integration of either model learner may be termed a *competent LCS*. That is, it is able to solve boundedly difficult problems—those with a minimal order complexity of k_m—effectively. The next chapter provides further evidence for the generality of the model-learning integration, investigating XCS's performance in several Boolean function problems including noisy and very large problems.

8

XCS in Binary Classification Problems

With all parts of the facetwise theory in place, this chapter applies XCS to several further binary classification problems experimentally, confirming the usefulness of the introduced XCS enhancements as well as the theoretic learning behavior. In particular, we investigate further the effects of tournament selection, fitness pressure, the influence of noise, niching, model-based offspring generation and overlapping problems. [1]

8.1 Multiplexer Problem Analyses

The multiplexer problem is traditionally studied in the LCS literature due to its interesting function properties. Appendix C gives an exact definition of the problem. As we saw in Chapter 5, the problem initially does not provide very strong fitness pressure. Even more severe is that the fitness pressure initially suggests the specialization of the value bits instead of the address bits since only the value bits result in a decrease in average error (information gain). This phenomenon is illustrated in Figure 8.1 (left-hand side) confirming the initial faster rise in specificity in the value attributes before the specificity in the address attributes takes over.

This section investigates fitness guidance in the multiplexer problem as well as performance in large multiplexer instances including the 70-multiplexer and the layered 135-multiplexer. Next, we investigate very noisy multiplexer problem instances. Finally, we compare XCS's performance with the performance of XCS with the BOA model building enhancement.

---

[1] Throughout this chapter, if not stated differently, parameters are set as follows: $N = 2000$, $\beta = 0.2$, $\alpha = 1$, $\varepsilon_0 = 10$, $\nu = 5$, $\theta_{GA} = 25$, $\chi = 1.0$, $\mu = 0.01$, $\gamma = 0.9$, $\theta_{del} = 20$, $\delta = 0.1$, $\theta_{sub} = 20$, and $P_\# = 0.8$.

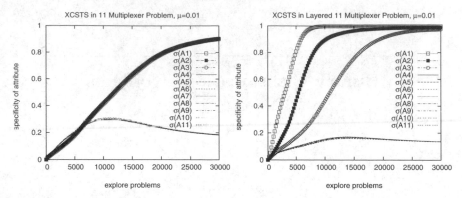

Fig. 8.1. Plotting the specificity evolution in the 11-multiplexer, value bits initially gain more specificity indicating the initial misleading nature of the problem. In the layered 11-multiplexer on the other hand, additional fitness guidance biases the fitness pressure towards the specialization of the address bits from the beginning.

8.1.1 Large Multiplexer Problems

XCS was shown to be able to solve the 70-multiplexer even with proportionate selection (Butz, Kovacs, Lanzi, & Wilson, 2004). However, in Butz, Goldberg, and Tharakunnel (2003) it was shown that XCSTS solves the problem much faster and much more reliably. XCS reached 100% performance in approximately 4 000 000 learning iterations using a population size of $N = 50 000$, a mutation rate $\mu = 0.04$ (niche mutation), and a sufficiently high initial specificity in the population ($P_\# = 0.75$).

Figure 8.2 shows the performance of XCSTS in the 70-multiplexer problem with a population size $N = 20k$, $N = 30k$, $N = 40k$, a mutation rate of $\mu = 0.01$, and an initially completely general population ($P_\# = 1.0$). The curves are averaged over 25 experiments. The graphs show that XCSTS with a population size of $40k$ solves the problem within 1 500 000 learning iterations. Decreasing the population size to $20k$ results in a much harder problem and all runs except one converged after 3 500 000 problems. The last run took more than 5 000 000 problems to find the optimal solution. Due to the small size of the population it is hard to give more accurate classifiers reproductive opportunities, often loosing the detected higher accurate classifiers. No XCS run with roulette wheel selection was able to solve the problem within 6 000 000 learning steps in the applied parameter settings. Although XCS with proportionate selection is also able to solve the 70-multiplexer problem, it relies on a larger population size, larger initial specificity and a higher mutation rate (Butz, Kovacs, Lanzi, & Wilson, 2004).

Figure 8.2 also compares XCSTS's performance with uniform crossover with XCSTS's performance with the BOA-based recombination mechanism (setting 18/10) introduced in the last chapter. Due to the intelligent recombination, performance becomes more reliable so that 100% accurracy is reached

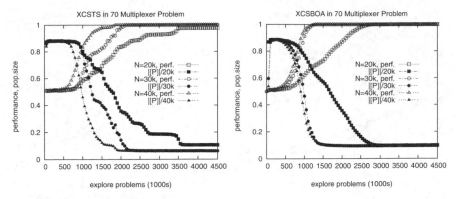

Fig. 8.2. XCSTS reliably solves the very large 70-multiplexer problem. A smaller population size delays the learning progress. Substituting uniform crossover with the Bayesian network-based recombination approach results in more effective genetic search and thus faster and more reliable learning.

in all runs. The comparison also points out the general suitability of the XCS/BOA combination.

In contrast to the multiplexer problem, the layered multiplexer problem provides strong fitness guidance, leading to a domino-like convergence of specificities. In fact, each specialization of an address bit cuts the consequential reinforcement range in half, strongly decreasing the expected average error. Thus, the layered multiplexer problem provides strong fitness guidance towards specializing the address bits. This observation is confirmed when plotting the specificity progress of each attribute in the 11-multiplexer problem. Figure 8.1 (right-hand side) shows how the specificities of each attribute increase progressively. The value bits are less misleading than in the normal multiplexer (left-hand side) and gain specificity slower than the more important address bits. This fitness guidance is certainly very helpful for the XCS learning mechanism.

Figure 8.3 shows the performance of XCSTS in the layered 70 and 135-multiplexer problems. Comparing the performance in the layered 70-multiplexer problem with that of the normal 70-multiplexer problem in Figure 8.2, we see that XCSTS is able to solve the 70-multiplexer problem much faster if reward is layered, confirming the successful exploitation of the available fitness guidance. With this additional fitness guidance, XCSTS is also able to solve the layered 135-multiplexer problem successfully. Due to the additional fitness guidance, the evolutionary process is much more directed causing less additional specializations of unnecessary attributes, effectively decreasing the specificity in the value attributes. The specificity decrease lowers the reproductive opportunity bound, allowing the evolution of a solution with a population size as low as $N = 20k$.

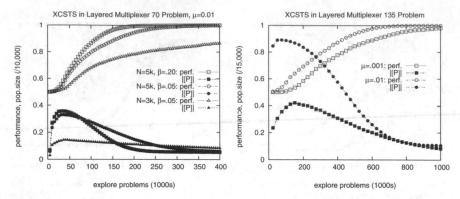

Fig. 8.3. Due to the additional fitness guidance available in the layered multiplexer problem, XCS is able to solve the 70 (left-hand side) as well as the 135 (right-hand side) problem much faster than the corresponding normal multiplexer problem instances.

8.1.2 Very Noisy Problems

When adding Gaussian noise with a large standard deviation ς, XCSTS also has a hard time evolving an accurate population. Figure 8.4a) shows that the speed of learning is strongly decreased when a noise of $\varsigma = 500$ is added. Perfect performance can only be reached if the learning rate β is lowered, effectively decreasing the noise in the parameter estimations of the classifiers. Even higher noise values prevent XCSTS from reaching 100% knowledge reliably (Figure 8.4b). When setting $\varsigma = 700$, hardly any learning is observable.

It is possible to determine the actual expected standard deviation of classifiers in problems with additional Gaussian noise. Given two normal distributions $N_1(\mu_1, \varsigma_1)$ and $N_2(\mu_2, \varsigma_2)$ and noting that a classifier approximates the mean average deviation (MAD), a classifier that is only applicable in situation-classification combinations that yield payoff distributed by N_1 will have approximately a prediction error estimate of $0.8\varsigma_1$ (since $MAD(N(\mu, \varsigma)) = \sqrt{2/\pi}\varsigma \approx 0.8\varsigma$). The prediction error estimate of a classifier that encounters the first payoff distribution with probability p and the second payoff distribution with probability $(1 - p)$ can then be approximated as follows:

$$\varsigma^2 = E(X^2) - E(X)^2 =$$
$$= pE(X_1^2) + (1 - p)E(X_2^2) - (p\mu_1 + (1 - p)\mu_2)^2$$
$$= pE(X_1^2) + (1 - p)E(X_2^2) - (p\mu_1)^2 - ((1 - p)\mu_2)^2 - 2p(1 - p)\mu_1\mu_2$$
$$= p(E(X_1^2) - \mu_1^2) + p\mu_1^2 - (p\mu_1)^2 + (1 - p)(E(X_2^2) - \mu_2^2) + (1 - p)\mu_2^2 -$$
$$((1 - p)\mu_2)^2 - 2p(1 - p)\mu_1\mu_2$$
$$= p\varsigma_1^2 + (1 - p)\varsigma_2^2 + p(1 - p)\mu_1^2 + p(1 - p)\mu_2^2 - 2p(1 - p)\mu_1\mu_2$$
$$= p\varsigma_1^2 + (1 - p)\varsigma_2^2 + p(1 - p)(\mu_1 - \mu_2)^2 \tag{8.1}$$

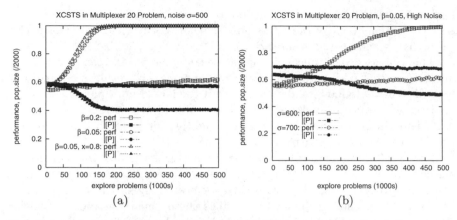

Fig. 8.4. (a) A lower learning rate β helps to derive accurate-enough fitness estimates in highly noisy problems. Crossover appears slightly disruptive. (b) XCSTS is able to solve the Boolean multiplexer problem up to a noise level of $\varsigma = 600$.

This prediction error approximation enables us to estimate the difficulty of the problem and to predict if a problem is still solvable at all. In the case of $\varsigma = 600$, the completely general classifier, for example, would experience a mean of 500 and a standard deviation of $(0.5 \cdot 600^2 + 0.5 \cdot 600^2 + 0.25 \cdot 1000^2)^{0.5} = 781$ so that the problem appears to be still solvable theoretically since the maximally accurate classifier experiences a standard deviation of $\varsigma = 600$. It also shows how much more difficult the problem is than the problem without any noise. Without any noise, accurate classifiers experience a standard deviation of $\varsigma = 0$ while the completely general classifiers experience a standard deviation of $\varsigma = 500$. The smaller the differences in the experienced standard deviations, the more the evolutionary process will be misled since the deviations are only approximated by the means of temporal difference learning techniques. Even stronger approximation mistakes are expectable and even inevitable in "young" classifiers that experienced only few parameter updates so far.

8.2 The xy-Biased Multiplexer

The xy-biased multiplexer problem was introduced elsewhere (Butz, Kovacs, Lanzi, & Wilson, 2001; Butz, Kovacs, Lanzi, & Wilson, 2004) to investigate fitness guidance. The problem combines the difficulty of the multiplexer problem iteratively. A first multiplexer function with x address bits chooses the y multiplexer that decides on the current class of the problem. Again, the problem is somewhat hierarchical but not as clearcut as the hierarchical problems we introduced earlier. In fact, the xy-biased multiplexer would be easy to solve by a typical hierarchical clustering approach (Duda, Hart, & Stork, 2001) in that the first x-bits partition the space focusing on the currently responsible classifier. However, learning the partition online is very hard since a specialization of one of the x bits does not necessarily yield any information gain (that is, increase in accuracy).

The xy-biased multiplexer is biased because the y multiplexer is slightly modified in that if all address bits are zero (or one) the result is a zero (one) regardless of the value bits, dependent on if the biased multiplexer is zero (or one) biased, respectively. This biases the problem in that the specialization of an address bit can increase accuracy slightly. Further information on the problem can be found in Appendix C.

Figure 8.5 shows performance curves in various larger instances of the xy-biased multiplexer problem. Curves are averaged over 20 runs and population size is set to $N = 15k$. To get an idea of how complex the final solutions are we use the function $||[O]||$ that specifies the optimal classifier population (Kovacs & Kerber, 2001). The measure defines the complexity of a problem as the size of the minimal, accurate, non-overlapping population that covers all environmental niches accurately. We note that $||[O]||(5, 1) = 192(l = 69)$, $||[O]||(4, 2) = 224(l = 84)$, $||[O]||(3, 3) = 244(l = 83)$, $||[O]||(2, 4) = 248(l = 78)$, and $||[O]||(1, 5) = 252(l = 63)$. The $[O]$-measure suggests that $(5, 1)$ is the simplest problem, while $(1, 5)$ is the most difficult one. The plots in Figure 8.5a show that the $(5, 1)$ setting is very hard to solve. The $(5, 1)$ is not learned after $1.5M$ learning steps. In the BOA setting (Figure 8.5c), however, $(5, 1)$ is solved the fastest, indicating disruptive effects due to uniform crossover. The problem length l also does not reflect problem complexity appropriately.

Performances indicate that in the x,y-biased multiplexer problem problem problem structure and fitness guidance is the main factor for a fast and reliable development of an accurate problem solution. For example, in the $(5, 1)$ problem performance reaches the 75% level very fast but has a hard time to evolve a maximally accurate solution. The challenge in the $(5, 1)$ problem is that the minimal order of difficulty is larger than one since the first bit of the five address bits is easily detected, but a specialization of any or even all of the other address bits does not increase accuracy before not at least one value bit is correctly set.

In the simpler problem instances, XCS with the Bayesian search is outperformed by normal XCS recombination. The problem is simpler with respect to evolutionary search but becomes harder with respect to initial accuracy. The extreme initial gain in accuracy in the $(5, 1)$ problem can cause disruption in the later learning progress. The Bayesian approach detects relevant dependencies much more effectively and alleviates the disruptive effects of simple, uniform crossover.

8.3 Count Ones Problems

The count ones problem was introduced in Butz, Goldberg, and Tharakunnel (2003) in order to investigate the benefits of recombination in the XCS

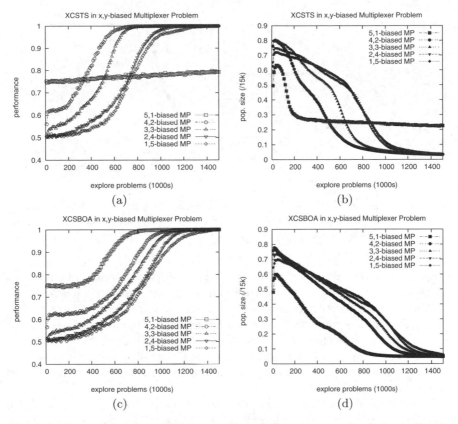

Fig. 8.5. Larger biased multiplexer problem instances are particularly challenging since the minimal order of problem difficulty increases. The Bayesian search approach can detect and propagate lower-level dependency structures more effectively.

classifier system. In fact, the count ones and especially the layered count ones problem can be compared to the one-max problem in genetic algorithms in that each relevant attribute increases fitness, that is accuracy, independent of each other. Further details on the problem can be found in Appendix C.

Investigating the specificity behavior of XCS classifiers in the layered 10/5 count ones problem (referring to the problem length $l = 10$ and the number of relevant attributes $k = 5$ in the problem), Figure 8.6 shows that the specificity in all relevant attributes behaves similarly while the specificity in all other attributes remains similarly low. In the count ones problem (left-hand graph), specificity reaches a level slightly above 3/5 since the optimal solution requires the specialization of three of the five attributes. The specificity stays slightly above the 3/5 because of the continuous specialization pressure caused by the application of the free mutation operator. In the layered count ones problem, 100% specificity is reached since all five attributes need to be specified to

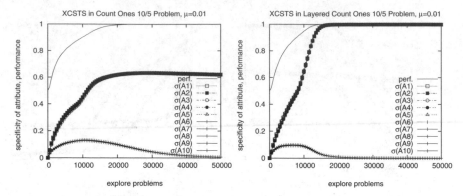

Fig. 8.6. The specificity curves in the 10/5 count ones problem (left-hand side) and the 10/5 layered count ones problem (right-hand side) exhibit the strong fitness guidance towards specializing the relevant attributes.

predict the resulting payoff level accurately. The performance line indicates that specificity converges shortly after complete performance is reached.

When comparing tournament selection with proportionate selection, Figure 8.7 shows once more that tournament selection is able to exploit fitness guidance much more effectively. Especially in the layered count ones problem, XCSTS solves the problem much more robustly. Additionally, crossover has a strong influence on performance since uniform crossover is very effective in the count ones problems.

Finally, we ran XCSTS in larger count ones problems with a problem length of $l = 100$ and seven relevant variables. Figure 8.8 shows that a lower mutation rate is mandatory to solve the problems effectively. Thus, XCS with roulette wheel selection fails to solve these larger instances since the necessary lower mutation rate is not sufficient to overcome the continuous generalization pressure. With tournament selection, however, the fitness pressure alone is strong enough and a complete problem solution is evolved. However, evolution is only successful, if crossover is applied. Mutation alone may still solve the problem but performance is strongly delayed. Due to successful recombinatory events combining detected substructures, the large chunks of seven bits are detected more efficiently.

8.4 Summary and Conclusions

This chapter has shown that XCS is applicable in a wide variety of classification problems. The problems range from non-overlapping to overlapping solution representations as well as from equally sized solution subspaces to highly unequally sized ones. Additionally, the problems were shown to possibly only provide noisy reinforcement feedback. XCS was able to solve all these

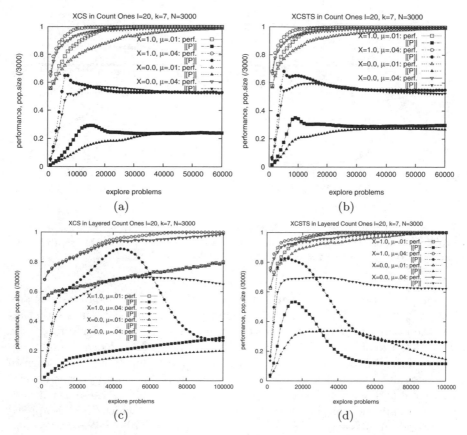

Fig. 8.7. Performance results in the 20/7 count ones (a,b) and layered count ones (c,d) of XCS with proportionate selection (a,c) and XCSTS (b,d) once again confirm the superiority of tournament selection. Additionally, recombination is very beneficial in these problems.

problems with an appropriate setting of population size and mutation rate. All other parameters did not need to be adjusted as long as starting out with a sufficiently general population controlled by the $P_\#$ parameter.

This confirms that despite the large number of parameters in the XCS system, most parameters have only a slight influence on performance and can be set to their standard values. Only population size and mutation rate may need to be set appropriately, dependent on the complexity of the problem. Thus, a future research goal is to adapt population size as well as mutation rate on the fly during a run when performance stalls due to over-specialized classifiers and a too small population size.

In conclusion, this section confirmed that XCS is capable of solving a large variety of Boolean function problems as proposed by the derived facetwise theory in the last chapters. The reader needs to be reminded that XCS, however,

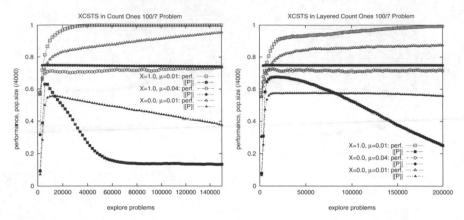

Fig. 8.8. In the larger 100/7 count ones (left-hand side) and layered count ones problems (right-hand side), a low mutation rate is necessary to enable reproductive opportunities. Population size is set to $N = 3000$.

is neither restricted to solve such Boolean function problems nor particularly designed to solve these Boolean function problems. In fact, in all the undertaken experiments feedback is solely provided in terms of reinforcement instead of direct, supervised feedback. Additionally, XCS learns online from one problem instance at a time. Thus, XCS is very much suited to learn in more cognitive scenarios in which feedback is available only in terms of reinforcement and learning is required to be undertaken online. These conclusions are also confirmed in the two subsequent chapters in which XCS is applied to several (multistep) MDP problems in which reward backpropagation is necessary, as well as in a variety of real-world data problems in which the handling of real-valued and nominal attributes is required. Additionally, in both scenarios problem instance sampling is not uniform over the problem space.

9

XCS in Multi-Valued Problems

This chapter investigates XCS's performance in various real- and/or nominal valued datasets as well as in function approximation problems. The application to problems other than binary valued ones requires a modification of the XCS classifier system condition parts as well as its genetic operators including covering, mutation, and crossover.

Besides the representational and operator enhancements, we revisit the established facetwise XCS theory in the real-valued and nominal problem domain. Essentially, it is necessary to revise the specificity definition for the new representation of classifier condition parts. Also the frequency of niche occurrence needs to be redefined due to the real-valued features and should be based on volume times sampling distribution.

Our performance analysis shows that XCS is able to learn competitively in various datasets taken from the University of California at Irvine (UCI), machine learning repository (S. Hettich & Merz, 1998) and other sources. In the analysis, we compare XCS's performance with many other established machine learning systems.

The main objective of this chapter is to confirm that XCS is a valuable learning system. The reader should keep in mind, though, that XCS is an online learning RL system and not a pure datamining system. Thus, the competitive results confirm XCS's learning competence. The real potential of the XCS mechanism, however, may lie rather in more reinforcement-based or predictive, online learning tasks as addressed in the subsequent chapter.

This chapter first introduces the necessary enhancements to XCS to be able to handle nominal values, real values, as well as a mixture of both. Section 9.2 outlines the consequently necessary theory modifications to ensure covering, schema supply, reproductive opportunities, and solution support in real-valued domains. Next, we investigate XCS's performance in a large set of datamining problems. Finally, we enhance XCS's predictive capabilities to linear predictions and show its capabilities in function approximation problems. Summary and conclusions discuss the results from a broader perspective.

9.1 XCS with Hyperrectangular Conditions

XCS was first applied to real-valued problems in Wilson (2000). The modified system, often referred to as XCSR, underwent changes in its condition representation as well as the operation of mutation and covering to be able to handle real-valued problem instances. Stone and Bull (2003) showed that the originally proposed condition structure that specifies the condition center and its width in each dimension (often referred to as center-spread method) biases the learner towards learning solution value boundaries, which is not desirable in the general case. Consequently we start by coding a condition part as a conjunction of intervals, specifying the lower and upper bound of each interval, as described in Wilson (2001b). The condition effectively specifies a hyperrectangle in which the classifier is applicable.

In the hyperrectangular case, the condition part consists of a conjunction of intervals represented by upper and lower boundaries (l_i, u_i), that is, $C = (l_1, u_1, l_2, u_2, ..., l_l, u_l)$. A condition matches a current real-valued problem instance $s \in S$ if the problem instance lies within all intervals specified by the classifier condition. Thus, the prediction depends on the resulting hyperrectangular problem space specified in the conditions.

Mutation changes the lower or upper bound with probability μ. If mutation is applied, the bound is increased or decreased by a value uniformly randomly picked between zero and m_0. In Wilson's original paper parameter m_0 was an absolute value equal in all problem features. To be more independent of the problem domain, we chose to define m_0 as the relative fraction of the feature range. For example, if a feature may take values between 0 and 10 and $m_0 = 0.2$, then the mutation interval ranges between zero and two. If mutation mutates an attribute below the lower value boundary of the problem or above the upper value boundary, the new bound is set to the lower or upper boundary, respectively. If mutation mutates the lower boundary above the upper boundary or vice versa, the boundary values are swapped.

Alternatively, mutation may also be defined relative to the current interval covered by the condition. In this case, mutation may increase or decrease the specified interval by a certain percentage of the whole interval (set to 50 percent in the subsequent experiments).

The covering operator is defined similarly to the binary covering operator. Given a currently uncovered problem instance, each feature is covered generating an interval for which the lower bound is chosen uniformly randomly between zero and r_0 distance from the current value and the upper bound is similarly chosen above the current value. Although r_0 could also be defined relative to the problem domain as is done for mutation, due to its importance solely in the beginning of a run we chose not to do so.

Note that the enhancements are directly taken from Wilson's publication (except for the relative mutation). Certainly, other types of mutation such as a Gaussian mutation approach as used in evolution strategies (Rechenberg,

1973; Bäck & Schwefel, 1995) might result in additional learning performance improvement. However, such enhancements are left for future research work.

In the case of nominal values, we apply the same enhancements, in that we convert the nominal values into an integer representation. The operators are then applied similarly to the real-valued case only that mutation, if applied, increases or decreases the boundary by at least one unit. In the case of only two nominal values, mutation is applied equivalently to the binary case.

9.2 Theory Modifications

Facing a new representation, the notion of niche occurrence, representatives, and specificity needs to be redefined. We are making use of the notion of volume to define problem subspaces as well as specificity. To keep the notation simple, we assume that each real-valued attribute ranges between zero and one.

With this constraint, we can define the volume of a classifier condition C by

$$\texttt{Vol}(C) = \prod_{i=1}^{l}(u_i - l_i).\tag{9.1}$$

The volume effectively defines the size of the subspace in which the classifier matches. Assuming a uniform random distribution of problem instances over the whole problem space, the definition of volume coincides with the probability of matching a problem instance. Consequently, the completely general classifier, in which all upper boundaries are equal to one and all lower boundaries are equal to zero, has a volume of one.

9.2.1 Evolutionary Pressure Adjustment

With respect to the previous specificity equation, few adjustments are necessary. Set pressure applies similarly to the binary case. Assuming a Gaussian distribution over specificities (which is exactly the case when sampling uniformly randomly and ignoring boundary effects), the resulting expected specificity in an action set (match set) is determined by multiplying the Gaussian with the probability of matching (which is equal to one minus the generality, which is equal to one minus the volume). Thus, the specificity in an action set can be expected to be similarly decreased.

Mutation in the used definition, however, does not have any effect on specificity when disregarding boundary effects. Since the increase of a boundary is as likely as the decrease of a boundary, the overall interval size is expected to stay the same. However, if a boundary is very close to the lower bound (upper bound) or essentially equal, mutation has a tendency to specialize since further generalization has no effect. Similarly, if the current interval is very small, mutation tends to generalize. Stone and Bull (2003) investigated a similar phenomenon, focusing on the effect on the boundary representation. As in the binary case, crossover has no effect on generality in the real-valued case.

In sum, the previously analyzed set pressure is also present in real-valued problems. Since mutation does not have an impact on specificity except when close to the boundaries, XCS can be expected to evolve fairly general classifiers as long as no fitness pressure applies.

Subsumption propagates syntactically more general classifiers, as in the binary case. Since especially in datamining problems problem instances are usually not uniformly distributed over the problem space, the semantic generalization pressure (that is, the set pressure) does not apply as prominently as in the binary problems investigated. Thus, subsumption pressure may have stronger generalization impacts as long as the problem allows the evolution of completely accurate classifiers.

9.2.2 Population Initialization: Covering Bound

With the notion of volume, we can derive the probability that a covering classifier exists, given a certain population size N and the covering operator r_0. In particular, a randomly generated classifier has an average volume of r_0^l since the interval starts on average half r_0 below the current value and stops half r_0 above the current value disregarding boundary effects. Note that if a value were to be circular, in that the lower value equates the highest value, then there would be no boundary effects and r_0 would be the exact expectable interval.

Classifiers in a randomly initialized population can be consequently expected to match with probability r_0^l. Similar to the covering bound in binary domains, we can derive a covering bound for real-valued domains as

$$P(\text{cover}_r) = 1 - (1 - r_0^l)^N. \qquad (9.2)$$

To ensure covering in real-valued domains, operator r_0 should consequently be set high enough. For example, requiring a certain confidence θ that a current problem instance is covered in the population given a fixed population size N, r_0 should be set as

$$r_0 > \sqrt[l]{1 - \sqrt[N]{1 - \theta}}. \qquad (9.3)$$

Due to boundary effects, some intervals generated during covering can be expected to be smaller so that the actual value of r_0 should be chosen slightly larger.

9.2.3 Schema Supply and Growth: Schema and Reproductive Opportunity Bound

To derive the schema and reproductive opportunity bound for the real-valued domain, it is necessary to re-define a representative of a certain problem niche.

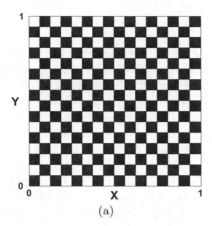
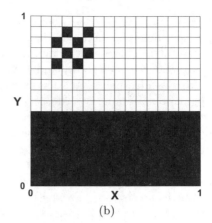

(a)   (b)

Fig. 9.1. (a) In a highly scattered checker board problem, it becomes extremely hard for XCS to evolve accurate classifiers when starting with overgeneral ones. (b) Starting from the overgeneral side, XCS faces a challenge if the problem space is highly unequally distributed and overgeneral classifiers prevent the evolution of more specialized classifiers.

Our intuitive definition in Chapter 6 may carry over in that a classifier may be considered a representative if it is at least as specific as the niche it represents. On the other hand, a slightly more general classifier may also be considered a representative if it does not overlap with any problem instance that belongs to another class. In general, since decision boundaries are never as clearcut as in the binary domain, the definition of a representative becomes less exact.

Additionally, the minimal order of problem difficulty k_m is not definable. In fact, it appears to vanish since the real-valued representation allows any kind of overlap so that the class distribution of a classifier cannot be approximated as easily. The maximally difficult problem appears to be a highly separated checker-board problem as illustrated in Figure 9.1a. As long as the covering operator is not chosen sufficiently specific, XCS can be expected to get stuck with an overgeneral, highly inaccurate representation since there is no specialization pressure towards evolving a more distributed problem representation unless a near-exact subsolution is identified (that is, one small rectangle in the checker problem).

The problem becomes even more clearly pronounced in the problem shown in Figure 9.1b. Starting for example with maximally general classifiers, XCS can be expected to evolve the shown class boundary $y = 7/16 = .4375$. However, the small additional positive cases in the upper left part of the problem space are very hard to identify and separate from the rest of the problem space. In the shown problem, the classifier that considers the complete space x and the space above $y = .4375$ is more accurate than any other classifier that specifies a subspace above $y = .4375$, and comprises the positive cases in the upper left except for a classifier that singles out the positive cases (rectangles). Thus, fitness guidance is missing and the problem's minimal order specificity k_m becomes apparent.

In this problem case, default hierarchies, which were suggested a long time ago for LCS systems (Holland, 1975), may do the job. Since the accuracy is high in a classifier that identifies the upper part of the problem space, it might be maintained as the default rule. However, since it is not always correct, classifiers that identify subspaces and particularly incorrectly specified problem cases may be able to solve the problem completely accurately. Another option would be to induce highly specialized classifiers in exceptional cases in which a general near-accurate classifier has an incorrect prediction. However, an inappropriate hunt for a classification of incorrect, noisy problem instances needs to be avoided. Given knowledge of the amount of noise in a problem, it would be possible to apply more directed search operators. In the general reinforcement-based learning case, however, such special operators can be expected to do more harm than result in any performance improvement.

9.2.4 Solution Sustenance: Niche Frequencies

Once an accurate representation evolved, the niche bound carries over rather directly. Given a niche of a certain volume and given further the maximally accurate classifier that covers that niche, the classifier will undergo the same Markov-chain reproduction and deletion process identified in Chapter 6. Thus, similar to the binary case, we can define the problem difficulty as the volume of the smallest problem niche. Since the volume equates the probability of niche occurrence p (as long as problem instances are uniformly randomly sampled from the whole problem space), the population size bound with respect to the niche bound is identical to the one derived for the binary case.

9.2.5 Obliqueness

Certainly there is an additional problem with respect to real-valued problem domains, which is the problem of obliqueness. Our current representation of condition parts does not allow us to represent oblique decision boundaries. XCS approximates an oblique boundary using a piece-wise approach, separating the problem space into small hyperrectangles. Again, the system faces the problem of reproductive opportunities and overgenerality. Since the boundaries are oblique, specializations may not result directly in an improvement of accuracy. If a more specialized classifier was found, the competition with the slightly less accurate classifier is a challenging one requiring several evaluations before the superiority of the more specialized classifier is detected. In effect, the more accurate classifier is likely to be deleted before undergoing any reproductive events.

The exact formulation of this problem and the resulting population size requirements are not investigated any further herein. Instead, the next section investigates XCS's performance in its current form. The results confirm that even in its current form, XCS shows able to solve datamining problems machine learning competitively.

9.3 Datamining

Due to the variable properties of the investigated datasets including real values, nominal values, and binary features, we use a hybrid XCS/XCSR approach that can handle any feature combination—as done elsewhere before (Bernadó, Llorà, & Garrell, 2002; Bernadó-Mansilla & Garrell-Guiu, 2003). In essence, each condition attribute is handled separately during matching, mutation, and covering, dependent on the type of problem feature.

The datamining analysis has two main objectives: (1) to compare XCS's performance of tournament selection with that of proportionate selection; (2) to compare XCSTS's performance with that of other machine learning algorithms. We first introduce the datasets that we investigated. Next, we present the results.

9.3.1 Datasets

In Table 9.1 we show the datasets we have selected including datasets from the UCI machine learning repository (S. Hettich & Merz, 1998) as well as a few other datasets. The other datasets are the `led.noise10` dataset which codes the seven lines of an LED display in binary. An additional 10% noise is added to the instances (features and class) to evaluate the approximation of the algorithm. The set was frequently used in the literature to investigate the learning performance on a dataset with added artificial noise. It was shown that the maximal performance achievable with 10% noise is 75% accuracy (Llorà & Goldberg, 2002). Secondly, the `tao` problem was previously investigated with the XCS system and other learning systems (Bernadó, Llorà, & Garrell, 2002; Bernadó-Mansilla & Garrell-Guiu, 2003). The problem is a dataset sampling uniformly randomly from the tao figure where white areas are assigned class zero and black areas class one. The dataset consequently has only oblique decision boundaries making it hard to learn with linear separators such as the interval coding we use in the XCS system.

Table 9.1 shows that the investigated datasets have various properties consisting of real, integer, nominal, and binary features. Hereby, nominal and integer features that only have two values are counted as binary features. Each dataset has a different number of features, of problem instances, and of solution classes. Also the number and distribution of instances per class vary. Finally, the number of missing values in the datasets vary.

To evaluate the performance of a learner, we apply stratified 10-fold cross-validation experiments (Mitchell, 1997).

With respect to the XCS application, we need to mention that XCS currently uses a simple strategy for missing values in a dataset, simply assuming

Table 9.1. The dataset properties indicate the number of problem instances (#Inst), the number of features (#Fea), of real-valued features (#Re), of integer-valued features (#In), of nominal features (#No), of binary features (#Bi), of classes (#Cl) as well as the percentage of instances belonging to the smallest class (%CMi) and the majority class (%CMa), of instances with missing values (%MVi), of features with missing values (%MVf), and of missing value values (#MVv).

Domain	#Inst.	#Fea	#Re	#In	#No	#Bi	#Cl	%CMi	%CMa	%MVi	%MVf	%MVv
anneal	898	38	6	0	13	19	5	0.9	76.2	0.0	0.0	0.0
audiology	226	69	0	0	8	61	24	0.4	25.2	98.2	10.1	2.0
autos	205	25	15	0	6	4	6	1.5	32.7	22.4	28.0	1.2
balance-scale	625	4	4	0	0	0	3	7.8	46.1	0.0	0.0	0.0
breast-cancer	286	9	0	0	6	3	2	29.7	70.3	3.1	22.2	0.3
breast-w	699	9	0	9	0	0	2	34.5	65.5	2.3	11.1	0.3
bupa	345	6	6	0	0	0	2	42.0	58.0	0.0	0.0	0.0
cmc	1473	9	2	0	4	3	3	22.6	42.7	0.0	0.0	0.0
colic	368	22	7	0	13	2	2	37.0	63.0	98.1	95.5	23.8
credit-a	690	15	6	0	5	4	2	44.5	55.5	5.4	46.7	0.6
credit-g	1000	20	7	0	11	2	2	30.0	70.0	0.0	0.0	0.0
diabetes	768	8	8	0	0	0	2	34.9	65.1	0.0	0.0	0.0
glass	214	9	9	0	0	0	6	4.2	35.5	0.0	0.0	0.0
heart-c	303	13	6	0	4	3	2	45.5	54.5	2.3	15.4	0.2
heart-c1	296	13	6	0	4	3	2	45.9	54.1	0.0	0.0	0.0
heart-h	294	13	6	0	4	3	2	36.1	63.9	99.7	69.2	20.5
heart-statlog	270	13	13	0	0	0	2	44.4	55.6	0.0	0.0	0.0
hepatitis	155	19	2	4	0	13	2	20.6	79.4	48.4	78.9	5.7
hypothyroid	3772	29	6	1	2	20	4	0.1	92.3	100.0	27.6	5.5
ionosphere	351	34	34	0	0	0	2	35.9	64.1	0.0	0.0	0.0
iris	150	4	4	0	0	0	3	33.3	33.3	0.0	0.0	0.0
kr-vs-kp	3196	36	0	0	2	34	2	47.8	52.2	0.0	0.0	0.0
labor	57	16	8	0	5	3	2	35.1	64.9	98.2	100.0	35.7
led.noise10	2000	7	0	0	0	7	10	8.9	10.9	0.0	0.0	0.0
lymph	148	18	0	3	6	9	4	1.4	54.7	0.0	0.0	0.0
mushroom	8124	22	0	0	18	4	2	48.2	51.8	30.5	4.5	1.4
new-thyroid	215	5	5	0	0	0	3	14.0	69.8	0.0	0.0	0.0
primary-tumor	339	17	0	0	3	14	21	0.3	24.8	61.1	29.4	3.9
segment	2310	19	19	0	0	0	7	14.3	14.3	0.0	0.0	0.0
sick	3772	29	6	1	2	20	2	6.1	93.9	100.0	27.6	5.5
sonar	208	60	60	0	0	0	2	46.6	53.4	0.0	0.0	0.0
soybean	683	35	0	0	19	16	19	1.2	13.5	17.7	97.1	9.8
splice	3190	60	0	0	60	0	3	24.0	51.9	0.0	0.0	0.0
tao	1888	2	2	0	0	0	2	50.0	50.0	0.0	0.0	0.0
vehicle	846	18	18	0	0	0	4	23.5	25.8	0.0	0.0	0.0
vote	435	16	0	0	0	16	2	38.6	61.4	46.7	100.0	5.6
vowel	990	13	10	0	1	2	11	9.1	9.1	0.0	0.0	0.0
waveform-5000	5000	40	40	0	0	0	3	33.1	33.8	0.0	0.0	0.0
wdbc	569	30	30	0	0	0	2	37.3	62.7	0.0	0.0	0.0
wine	178	13	13	0	0	0	3	27.0	39.9	0.0	0.0	0.0
wpbc	198	33	33	0	0	0	2	23.7	76.3	2.0	3.0	0.1
zoo	101	17	0	1	1	15	7	4.0	40.6	0.0	0.0	0.0

a match to a missing value. Other strategies might be superior to this one as also suggested by our recent datamining comparison with the Pittsburgh-style learning classifier system GAssist (Bacardit & Butz, 2004). Additionally, nominal values are coded as integers. This might lead to performance drawbacks since the resulting integer ordering might be inappropriate with respect to the problem. Other set-based approaches might be more appropriate but are not investigated further herein. It also needs to be remembered that we code nominal features with only two values identical to binary features in contrast to previous datamining investigations with XCS (Bernadó, Llorà, & Garrell, 2002; Bernadó-Mansilla & Garrell-Guiu, 2003).

9.3.2 Results

Table 9.2 compares XCS's performance with proportionate selection with XCS's performance with tournament selection (XCSTS). The table clearly confirms the superiority of tournament selection, proving initial faster learning as well as better convergence properties. Additionally, the results show that $100k$ learning steps may not be enough to reach optimal performance. Besides the different number of learning steps and the different initializations, we also tested XCS on two covering parameter settings, causing the initial population to be either very general ($r_0 = 100$) or very specific ($r_0 = 4$). It can be seen that a more general initialization is often advantageous, preventing over-specializations. However, in some datasets such as the tao problem, an initially more specialized population is advantageous. This observation confirms the facetwise theory extension from the last section. The oblique boundaries in the tao problem are very hard to improve upon once a general solution was learned. Starting with an initially fairly specific population assures that the oblique boundaries are estimated more accurately, consequently reaching and maintaining a higher performance level.

We compared XCSTS with a number of other machine learning programs including C4.5 (revision 8) (Quinlan, 1993), the Naive Bayes classifier (John & Langley, 1995), PART (Frank & Witten, 1998), the instance based learning algorithm with one and three nearest neighbor setting (Aha, Kibler, & Albert, 1991; Mitchell, 1997) and the support vector machine implementation SMO with polynomial kernels of order one and three and with radial basis kernels (Platt, 1998). To establish a performance baseline, we also provide results for the simple majority class algorithm which chooses the majority class as the classification throughout. Moreover, to provide another, more challenging baseline result, we also ran the simple 1-R classifier that learns one rule that is conditioned on the most significant problem features, ranking all features according to their train-set error and dividing continuous values straight-forwardly requiring a minimal number of six instances of the optimal class in the interval (Holte, 1993). All machine learning methods were run using the WEKA machine learning package (Witten & Frank, 2000).

Performance comparisons confirm that XCS performs competitively to these other machine learning algorithms (Tables 9.3 and 9.4). Statistics are based on eight stratified 10-fold crossvalidation runs (that is, 80 experiments in each dataset). XCSTS learned for $500k$ steps in the presented results. As can be expected given the number of different datasets, XCSTS outperforms the other learners in some datasets whereas it is outperformed in others. Interestingly, XCSTS is outperformed in the soybean dataset by all other learning mechanisms (except majority and 1-R). Considering the soybean properties, two observations can explain this finding: (1) The soybean dataset consists of

Table 9.2. Crossvalidation performance of XCS comparing tournament selection (TS) with proportionate selection (PS) in diverse datasets after $100k$ learning iterations. Initialization is either general ($r_0 = 100$) or specific ($r_0 = 4$). The last column shows results of longer XCSTS runs ($500k$ learning iterations) comparing the results with the shorter runs. The $\blacklozenge$ ($\lozenge$) and $\blacksquare$ ($\square$) symbols indicate in which problems the performance in the first or second column is significantly better (worse) than the other columns with a significance level of .99 and .95 (pairwise t-test), respectively. The final two lines summarize the performance scores reporting how many times the first or second column was better/worse in classification performance than the respective column.

Database	TS($r_0 = 100$)	TS($r_0 = 4$)	PS($r_0 = 100$)	PS($r_0 = 4$)	TS($r_0 = 100, t = 500k$)
anneal	91.2±2.7	91.7±2.9	92.9±2.7 $\lozenge$	91.4±3.2	97.7±2.0 $\lozenge$
audiology	73.8±12.6	74.1±12.8	73.5±12.6	73.9±12.8	79.6±12.3 $\lozenge$
autos	64.7±9.6	13.4±6.9 $\blacklozenge$	62.0±10.3 $\blacksquare$	14.7±7.3	71.2±9.9 $\lozenge$
balance-scale	84.6±3.3	82.0±3.5 $\blacklozenge$	83.6±4.0 $\blacksquare$	83.6±3.0 $\lozenge$	81.1±3.8 $\blacklozenge$
breast-cancer	71.8±6.6	68.5±8.0 $\blacklozenge$	72.6±5.3	69.0±7.6	70.1±8.0 $\blacksquare$
breast-w	96.2±2.2	96.5±1.9	96.1±2.4	96.2±2.1	95.9±2.3
bupa	67.2±7.9	58.7±8.5 $\blacklozenge$	65.8±7.7	56.0±8.2 $\blacklozenge$	67.1±7.5
cmc	50.1±4.7	53.6±3.6 $\lozenge$	43.3±5.7 $\blacklozenge$	53.5±3.7	52.4±3.6 $\lozenge$
colic	84.5±5.8	84.8±5.5	84.3±6.2	84.8±6.1	84.0±5.8
credit-a	86.2±3.9	85.3±4.1 $\blacksquare$	85.7±3.9	85.1±3.8	85.6±3.5
credit-g	71.4±3.8	72.5±3.1	71.8±3.6	71.4±2.9 $\blacklozenge$	70.9±4.3
diabetes	74.3±4.5	67.3±3.5 $\blacklozenge$	74.4±4.5	65.6±2.6 $\blacklozenge$	72.4±5.3 $\blacklozenge$
glass	70.6±8.2	70.7±8.4	54.5±9.7 $\blacklozenge$	70.8±9.7	71.8±8.9
heart-c1	77.7±6.8	68.9±8.6 $\blacklozenge$	77.7±7.5	65.6±8.9 $\blacklozenge$	76.5±7.9
heart-c	77.4±7.9	68.1±8.5 $\blacklozenge$	77.0±7.6	66.0±8.2	77.2±6.9
heart-h	79.4±7.7	70.8±6.9 $\blacklozenge$	80.8±7.1 $\square$	70.1±6.3	77.8±8.0
heart-statlog	77.3±7.6	67.9±9.0 $\blacklozenge$	77.9±7.1	68.0±8.3	75.3±8.1
hepatitis	81.8±8.6	80.3±7.7	79.6±9.3 $\blacksquare$	79.9±8.4	80.7±9.2
hypothyroid	98.7±0.7	98.9±0.7	98.1±1.3 $\blacklozenge$	98.5±0.7 $\blacklozenge$	99.5±0.4 $\lozenge$
ionosphere	90.7±5.3	57.1±6.8 $\blacklozenge$	89.7±4.8	57.2±6.1	90.1±4.7
iris	92.6±7.2	95.0±5.0 $\lozenge$	84.7±12.8 $\blacklozenge$	95.2±5.0	94.7±5.1 $\square$
kr-vs-kp	98.7±0.7	99.0±0.6 $\lozenge$	97.7±0.8 $\blacklozenge$	98.3±0.8 $\blacklozenge$	98.9±0.6 $\square$
labor	80.9±15.5	86.6±15.3 $\lozenge$	70.0±12.0 $\blacklozenge$	86.4±14.7	83.5±14.8
led.noise10	72.1±3.7	72.1±3.8	67.4±5.3 $\blacklozenge$	67.5±5.4 $\blacklozenge$	72.7±3.1
lymph	81.7±9.5	83.5±9.5	81.2±10.2	81.9±10.2	79.8±10.2
mushroom	99.8±0.4	100.0±0.0 $\lozenge$	99.1±0.5 $\blacklozenge$	100.0±0.0	100.0±0.0 $\lozenge$
new-thyroid	94.8±4.5	94.8±4.8	93.9±5.2	94.4±4.6	95.4±4.6
primary-tumor	39.7±7.5	39.9±7.3	39.4±7.7	40.2±8.0	39.8±8.5
segment	92.5±1.7	16.8±1.1 $\blacklozenge$	92.2±1.9	16.8±1.1	95.8±1.3 $\lozenge$
sick	94.2±1.0	98.4±0.6 $\lozenge$	93.9±0.1 $\blacklozenge$	98.2±0.7	94.9±1.7 $\lozenge$
sonar	77.3±8.1	81.6±7.9 $\lozenge$	75.0±8.6	80.7±9.7	77.9±8.0
soybean	76.1±5.9	85.1±3.8 $\lozenge$	70.3±6.6 $\blacklozenge$	82.9±4.2 $\blacklozenge$	85.1±4.4 $\lozenge$
splice	88.9±4.2	28.0±1.0 $\blacklozenge$	73.6±14.4 $\blacklozenge$	28.1±0.9 $\lozenge$	93.8±1.5 $\lozenge$
tao	84.3±4.1	94.4±1.6 $\lozenge$	82.1±4.7 $\blacklozenge$	93.8±1.6 $\blacklozenge$	86.4±4.5 $\lozenge$
vehicle	73.8±4.4	25.1±0.5 $\blacklozenge$	73.4±4.0	25.1±0.5	74.3±4.7
vote	95.6±3.2	95.8±3.1	96.3±2.8 $\lozenge$	96.1±3.0	95.7±3.1
vowel	56.1±7.5	33.3±4.4 $\blacklozenge$	51.8±6.9 $\blacklozenge$	33.8±4.9	66.0±6.6 $\lozenge$
waveform-5000	82.5±1.7	33.8±0.0 $\blacklozenge$	82.2±1.5	33.8±0.0	82.5±1.5
wdbc	95.8±2.6	92.9±3.3 $\blacklozenge$	95.3±2.7	91.4±4.2 $\blacklozenge$	96.0±2.5
wine	95.8±4.7	97.1±4.0 $\lozenge$	93.6±5.8 $\blacklozenge$	95.4±5.3 $\blacklozenge$	95.6±4.9
wpbc	73.0±8.4	23.7±1.8 $\blacklozenge$	73.0±8.7	23.8±2.0	74.3±8.9
zoo	94.6±6.9	93.1±6.9	92.2±7.0 $\blacklozenge$	92.9±7.5	95.1±6.1
score 99%		17/10	15/2	11/2	2/12
score 95%		18/10	18/3	11/2	3/14

Table 9.3. Comparing XCSTS's performance with other typical machine learning algorithms confirms its competitiveness. The ♦ (◊) and ■ (□) symbol indicate in which problems XCSTS performed significantly better (worse) on a significance level of .99 and .95 (pairwise t-test), respectively. The final two lines summarize the performance scores.

Database	XCSTS	Majority	Main Ind.	C4.5	Naive Bayes	PART
anneal	97.7±2.0	76.2±0.8 ♦	83.6±0.5 ♦	98.6±1.1	86.6±3.4 ♦	98.4±1.3
audiology	79.6±12.3	26.9⊥5.5 ♦	49.6±10.2 ♦	81.2±11.0	76.8±15.4	83.3±12.6 ◊
autos	71.2±9.9	32.8±2.1 ♦	61.4±10.1 ♦	82.0±8.7 ◊	56.8±9.9 ♦	74.6±8.5
balance-scale	81.1±3.8	46.1±0.2 ♦	57.9±4.0 ♦	77.9±4.0 ♦	90.5±1.8 ◊	82.4±4.0
breast-cancer	70.1±8.0	70.3±1.2	67.0±6.8	73.8±6.2 □	72.7±8.1	69.8±7.1
breast-w	95.9±2.3	65.5±0.3 ♦	91.9±2.9 ♦	94.5±2.6 ♦	96.0±2.1	94.7±2.4 ■
bupa	67.1±7.5	58.0±0.8 ♦	55.9±7.1 ♦	65.0±8.6	54.8±8.3 ♦	64.3±8.2
cmc	52.4±3.6	42.7±0.2 ♦	47.5±3.7 ♦	51.6±4.1	50.4±4.1	50.0±4.0 ♦
colic	84.0±5.8	63.1±0.8 ♦	81.5±6.3	85.2±5.4	78.3±6.4 ♦	84.4±5.4
credit-a	85.6±3.5	55.5±0.4 ♦	85.5±4.1	85.4±4.2	77.8±4.2 ♦	84.0±4.5
credit-g	70.9±4.3	70.0±0.0	66.0±3.4 ♦	71.1±3.6	75.1±3.7 ◊	70.0±3.9
diabetes	72.4±5.3	65.1±0.3 ♦	72.2±4.2	74.2±4.6	75.6±4.9 ◊	73.9±4.6
glass	71.8±8.9	35.6±1.4 ♦	56.6±9.4 ♦	67.4±8.8	48.1±8.2 ♦	68.3±8.5
heart-c1	76.5±7.9	54.1±0.9 ♦	73.7±7.2	76.4±8.2	83.3±6.0 ◊	78.3±7.2
heart-c	77.2±6.9	54.5±0.7 ♦	72.7±7.0 ■	77.2±6.9	83.5±6.2 ◊	78.1±7.1
heart-h	77.8±8.0	64.0±0.9 ♦	80.7±7.1	79.5±7.9	84.3±7.5 ◊	80.3±7.7
heart-statlog	75.3±8.1	55.6±0.0 ♦	72.0±8.0	77.8±8.0	83.9±6.4 ◊	77.4±7.3
hepatitis	80.7±9.2	79.4±1.5	82.5±7.0	79.0±9.2	83.6±9.2	79.8±7.8
hypothyroid	99.5±0.4	92.3±0.3 ♦	96.3±0.9 ♦	99.5±0.3	95.3±0.6 ♦	99.5±0.4
ionosphere	90.1±4.7	64.1±0.6 ♦	81.8±6.0 ♦	90.0±5.3	82.5±7.1 ♦	90.7±5.2
iris	94.7±5.1	33.3±0.0 ♦	93.3±5.9	94.8±5.9	95.5±5.1	94.4±6.4
kr-vs-kp	98.9±0.6	52.2±0.1 ♦	67.1±1.8 ♦	99.4±0.4 ◊	87.8±1.9 ♦	99.1±0.6
labor	83.5±14.8	64.7±3.1 ♦	72.3±14.1 ♦	79.2±14.7	94.1±8.7 ◊	78.1±15.2
led.noise10	72.7±3.1	10.8±0.1 ♦	19.5±1.0 ♦	74.8±2.8 ◊	75.5±2.8 ◊	74.1±3.0
lymph	79.8±10.2	55.0±2.9 ♦	74.8±11.9	77.7±11.1	83.0±8.6	76.9±10.0
mushroom	100.0±0.0	51.8±0.0 ♦	98.5±0.4 ♦	100.0±0.0	95.8±0.7 ♦	100.0±0.0
new-thyroid	95.4±4.6	69.8±1.6 ♦	91.4±6.5 ♦	92.5±6.3 ♦	97.0±3.3	94.0±5.5
primary-tumor	39.8±8.5	25.3±3.2 ♦	28.0±4.9 ♦	43.1±7.3	50.7±8.7 ◊	41.8±7.6
segment	95.8±1.3	14.3±0.0 ♦	64.4±2.6 ♦	96.8±1.1 ◊	80.1±1.8 ♦	96.5±1.2 ◊
sick	94.9±1.7	93.9±0.1 ♦	96.2±0.8 ◊	98.7±0.5 ◊	92.8±1.3 ♦	98.6±0.6 ◊
sonar	77.9±8.0	53.4±1.2 ♦	62.1±9.5 ♦	73.8±8.5	67.8±9.6 ♦	74.4±9.5
soybean	85.1±4.4	13.5±0.5 ♦	39.8±2.6 ♦	91.9±3.2 ◊	92.8±2.8 ◊	91.4±3.1 ◊
splice	93.8±1.5	51.9±0.1 ♦	63.4±1.6 ♦	94.2±1.3	95.5±1.3 ◊	92.6±1.4 ♦
tao	86.4±4.5	50.0±0.0 ♦	72.3±3.0 ♦	95.6±1.3 ◊	80.9±2.9 ♦	94.2±2.1 ◊
vehicle	74.3±4.7	25.6±0.3 ♦	52.8±4.7 ♦	72.4±4.1	45.1±4.7 ♦	72.0±4.0 ■
vote	95.7±3.1	61.4±0.4 ♦	95.6±3.1	96.5±2.8	90.1±4.5 ♦	96.0±3.0
vowel	66.0±6.6	9.1±0.0 ♦	33.5±4.5 ♦	80.1±3.6 ◊	63.2±4.9	78.5±4.0 ◊
waveform-5000	82.5±1.5	33.8±0.0 ♦	53.9±1.9 ♦	75.2±2.0 ♦	80.0±1.5 ♦	77.9±1.7 ◊
wdbc	96.0±2.5	62.7±0.4 ♦	88.5±3.5 ♦	93.5±3.2 ♦	93.4±3.5 ♦	94.1±3.0 ♦
wine	95.6±4.9	39.9±1.7 ♦	78.0±9.7 ♦	93.3±6.5	97.2±4.0	92.4±6.1 ♦
wpbc	74.3±8.9	76.3±1.8	68.8±6.1 ♦	72.7±8.6	67.1±10.4 ♦	74.2±9.0
zoo	95.1±6.1	41.5±5.9 ♦	43.1±4.9 ♦	93.0±6.8	95.7±6.0	93.7±6.7
score 99%		38/0	29/1	5/8	19/12	5/6
score 95%		38/0	30/1	5/9	19/12	7/6

lots of nominal features and the conversion to an integer representation may be unfavorable in this dataset; (2) Many instances have missing values in this problem. The missing value strategy currently implemented in XCS of simply assuming a match in a missing feature might be inappropriate in this dataset.

Table 9.4. Continued comparison of XCSTS with other machine learning algorithms. The ♦ (◇) and ■ (□) symbol indicate in which problems XCSTS performed significantly better (worse) on a significance level of .99 and .95 (pairwise t-test), respectively. The final two lines summarize the performance scores.

Database	XCSTS	Inst.b.1	Inst.b.3	SMO(poly.1)	SMO(poly.3)	SMO(radial)
anneal	97.7±2.0	99.1±1.0 ◇	97.2±1.6	97.2±1.5	99.2±0.8 ◇	91.9±1.8 ♦
audiology	79.6±12.3	77.4±9.4	71.1±12.4 ♦	84.4±11.7 ◇	82.3±12.3	46.8±10.1 ♦
autos	71.2±9.9	73.5±9.3	67 4±10.0	70.6±9.1	77.2±9.1 ◇	45.5±6.8 ♦
balance-scale	81.1±3.8	78.5±4.2 ♦	86.7±3.0 ◇	87.7±2.4 ◇	90.7±2.8 ◇	87.8±2.7 ◇
breast-cancer	70.1±8.0	68.6±7.8	73.7±5.1 □	70.1±7.3	66.8±7.6	70.3±1.2
breast-w	95.9±2.3	95.6±2.1	96.6±2.0	96.7±1.9	95.9±2.1	96.0±2.2
bupa	67.1±7.5	62.2±7.8 ■	62.5±7.6	57.9±1.5 ♦	59.1±3.4 ♦	58.0±0.8 ♦
cmc	52.4±3.6	44.3±3.8 ♦	46.7±3.6 ♦	48.9±3.6 ♦	49.1±3.7 ♦	42.8±1.1 ♦
colic	84.0±5.8	79.0±5.7 ♦	81.4±5.7	82.8±6.2	77.6±6.7 ♦	84.1±5.3
credit-a	85.6±3.5	81.6±4.7 ♦	85.0±4.1	84.9±3.8	81.8±4.6 ♦	85.5±4.1
credit-g	70.9±4.3	72.0±3.9	72.7±3.4	74.9±3.7 ◇	70.6±4.0	70.1±0.2
diabetes	72.4±5.3	70.4±4.4	73.9±4.5	77.0±4.3 ◇	76.9±4.1 ◇	65.1±0.3 ♦
glass	71.8±8.9	70.3±9.1	70.6±8.4	57.0±7.7 ♦	65.7±8.9 ♦	35.6±1.4 ♦
heart-c1	76.5±7.9	75.8±6.9	82.2±6.5 ◇	83.5±6.1 ◇	76.6±7.8	82.8±6.6 ◇
heart-c	77.2±6.9	75.7±7.7	82.3±6.5 ◇	83.4±6.5 ◇	77.9±6.8	83.2±6.3 ◇
heart-h	77.8±8.0	78.6±6.8	82.2±7.4 □	82.7±8.0 ◇	78.0±7.5	82.0±7.4 □
heart-statlog	75.3±8.1	75.1±7.5	79.2±7.2 □	83.8±6.3 ◇	79.0±6.5 ◇	82.9±6.5 ◇
hepatitis	80.7±9.2	80.9±8.3	80.2±8.6	85.6±8.7 □	82.7±7.9	79.4±1.5
hypothyroid	99.5±0.4	91.4±1.0 ♦	93.2±0.8 ♦	93.6±0.5 ♦	93.9±0.6 ♦	92.3±0.3 ♦
ionosphere	90.1±4.7	86.7±5.1 ♦	86.0±5.0 ♦	87.8±5.0	88.0±5.8	75.7±4.7 ♦
iris	94.7±5.1	95.4±5.2	95.2±5.4	96.3±4.6	92.9±6.0	90.6±9.0 ■
kr-vs-kp	98.9±0.6	90.5±1.6 ♦	96.5±1.1 ♦	95.8±1.2 ♦	99.6±0.4 ◇	91.4±1.6 ♦
labor	83.5±14.8	85.7±13.6	93.6±9.1 ◇	93.0±10.7 ◇	91.8±11.5 □	64.7±3.1 ♦
led.noise10	72.7±3.1	65.3±6.2 ♦	74.8±2.7 ◇	75.4±2.8 ◇	74.1±2.9	75.4±2.9 ◇
lymph	79.8±10.2	81.8±9.7	82.3±8.7	86.0±8.0 ◇	83.4±9.0	81.5±10.4
mushroom	100.0±0.0	100.0±0.0	100.0±0.0	100.0±0.0	100.0±0.0	99.8±0.1 ♦
new-thyroid	95.4±4.6	96.6±4.1	94.2±4.8	89.6±6.0 ♦	89.4±6.3 ♦	69.8±1.6 ♦
primary-tumor	39.8±8.5	34.3±7.5 ♦	46.0±7.6 ◇	47.9±7.8 ◇	42.3±7.2	25.3±3.2 ♦
segment	95.8±1.3	97.1±1.2 ◇	96.1±1.3	92.9±1.4 ♦	96.4±1.2	82.0±2.2 ♦
sick	94.9±1.7	96.1±0.8 ◇	96.1±0.8 ◇	93.9±0.1 ♦	95.9±0.7 ◇	93.9±0.1 ♦
sonar	77.9±8.0	86.4±7.0 ◇	83.5±7.7 ◇	77.5±9.1	85.1±8.0 ◇	68.6±7.4 ♦
soybean	85.1±4.4	90.2±3.1 ◇	91.1±3.0 ◇	92.7±2.7 ◇	93.2±2.7 ◇	87.9±2.6 ◇
splice	93.8±1.5	75.8±2.6 ♦	77.6±2.5 ♦	93.1±1.6	96.9±1.0 ◇	96.3±1.1 ◇
tao	86.4±4.5	96.6±1.2 ◇	96.1±1.3 ◇	84.0±2.7 ♦	84.3±2.7 ■	83.6±2.7 ♦
vehicle	74.3±4.7	69.4±4.0 ♦	70.0±4.4 ♦	74.4±4.1	83.0±3.7 ◇	41.7±3.8 ♦
vote	95.7±3.1	92.3±3.9 ♦	93.0±3.9 ♦	95.9±3.0	94.9±3.3	94.7±3.2
vowel	66.0±6.6	99.1±0.9 ◇	97.1±1.7 ◇	70.5±4.0 ◇	99.2±0.9 ◇	31.0±5.3 ♦
waveform-5000	82.5±1.5	73.5±1.7 ♦	77.6±1.6 ♦	86.4±1.4 ◇	81.5±1.6 ♦	85.4±1.5 ◇
wdbc	96.0±2.5	95.3±2.7	97.1±1.9 □	97.7±1.9 ◇	97.3±2.1 ◇	92.3±2.9 ♦
wine	95.6±4.9	95.2±4.4	96.7±3.9	98.6±2.6 ◇	97.2±3.6	41.3±3.2 ♦
wpbc	74.3±8.9	71.4±9.9	74.1±8.7	76.4±3.7	78.4±8.6	76.3±1.8
zoo	95.1±6.1	97.5±4.7	93.3±6.9	96.4±5.6	95.0±6.6	73.4±9.1 ♦
score 99%		13/7	9/11	9/17	8/13	23/8
score 95%		14/7	9/15	9/18	9/14	24/9

9.4 Function Approximation

Besides real-valued datamining problems, XCS can also be used to solve general function approximation problems. Wilson (2001a) showed that XCSR can be enhanced with linear prediction structures to learn piece-wise linear function approximations. XCS with hyperrectangular conditions was enhanced in the following way.

The prediction part of a classifier is now not represented by a constant but by a weight vector W of size $l + 1$. A reward prediction is now computed by the inner product of the prediction vector W and the input vector s. To normalize the state vector, we determine the distance from the lower bound of the condition part and enhance the resulting difference vector by an initial one (to account for the offset weight). The consequent prediction array entries are computed as follows:

$$P(A) = \frac{\sum_{cl.A=A \wedge cl \in [M]} ||cl.W(cl.l - s)^*||cl.F}{\sum_{cl.A=A \wedge cl \in [M]} cl.F}, \tag{9.4}$$

where the $*$ denotes the vector enhancement with an additional one entry.

As before, XCS iteratively learns from the successive problem instances and uses feedback ρ to update the prediction part. The basic delta update rule for the weight vector W is used, that is,

$$w_0 \leftarrow w_0 + \eta(\rho - ||R(l - s)^*||),$$
$$w_i \leftarrow w_i + \eta(\rho - ||R(l - s)^*||)(s_i - l_i), \tag{9.5}$$

where η denotes the learning rate. Note that this specification does not apply a normalization factor used elsewhere. The interested reader is referred to Wilson (2002) for alternative definitions.

Besides the enhancement of linear predictions, other condition structures may be more suitable for function approximation tasks. Essentially, hyperrectangular conditions assume axis-parallel nonlinearities. The following condition structure progressively alleviate this assumption.

9.4.1 Hyperellipsoidal Conditions

While the originally introduced hyperrectangular condition structures can only express axis-parallel partitions, the following hyperspheroidal and hyperellipsoidal condition structures increase the axis-independence of classifiers. Besides the introduction of such alternative condition structures, the subsequent positive results confirm that XCS may be combined with any condition structure that measures similarity. Thus, XCS's conditions may be expressed by any kernel structure (Shawe-Taylor & Cristianini, 2004). The following hyperspheres and hyperellipsoids are related to general Gaussian kernels.

Hyperspheres

To represent hyperspheres, the condition part is represented by a center point and a deviation, that is, $C = (m_1, m_2, ..., m_n, \sigma_*)$. A classifier is active if the current problem instance lies within a certain range of the specified hypersphere. We use the Gaussian kernel function to determine the activity of a classifier:

$$cl.ac = \exp\left(-\frac{||s - m||^2}{2\sigma_*^2}\right),$$ (9.6)

where $m = (m_1, ..., m_n)$. To determine if a classifier is active, its activity $cl.ac$ needs to lie above a threshold θ_m. The smaller θ_m, the larger the receptive field and thus the probability of matching of a condition structure. In a sense, θ_m is the pendant to σ_*. Since σ_* is evolved, θ_m can be fixed. It is set to $\theta_m = .7$ throughout the experiments. It should be noted that spheres could be defined slightly more effectively simply using the l^2 norm. The chosen radial bases conditions, emphasize the relation to kernels and the potential of substituting XCS conditions with any other kernel-based basis structure.

As in the case of hyperrectangles, covering, mutation and crossover need to be adjusted. When creating a covering classifier, the center is set directly to the current values (that is, $m = s$) and the deviation σ_* is set uniformly randomly between zero and r_0 (zero excluded). For mutation, we define a relative mutation similar to the one for hyperrectangles above. Each attribute in the condition part is mutated with a probability μ. If an attribute of the center is mutated, the new value m_i' is set to a value uniformly randomly chosen in the interval the classifier applies in, that is, $|m_i - m_i'| \leq \sigma_*\sqrt{-2\log\theta_m}$. The standard deviation σ_* is either increased or decreased (equally likely) between zero and 50% chosen uniformly randomly. If the standard deviation is larger than a deviation necessary to contain the whole problem space, it is set to that value, that is:

$$\sigma_* \leq \frac{\sqrt{\sum_{i=1}^n (u_i^* - l_i^*)^2}}{\sqrt{-2\log\theta_m}},$$ (9.7)

where u_i^* and l_i^* denote the maximum upper and lower values of each dimension, respectively. As crossover operator we chose uniform crossover since no dependencies between the dimensions are known. A classifier is considered as more general if its hypersphere completely contains the other hypersphere.

Hyperellipsoids

Clearly, the common radius in hyperspheres restricts the expressibility of the conditions. Hyperellipsoids allow different radii (or deviations) for different axes. A hypersphere is a special case of an hyperellipsoid. To represent hyperellipsoids with an XCS condition, the condition part is redefined as follows:

$$C = (m_1, m_2, ..., m_n, \sigma_1, ..., \sigma_n)$$ (9.8)

The condition is now represented by a center point and deviations in all n dimensions. The activity of a classifier given the current problem instance s is now defined as:

$$cl.ac = \exp\left(-\sum_{i=1}^{n} \frac{(s_i - m_i)^2}{2\sigma_i^2}\right), \tag{9.9}$$

effectively dividing in each dimension the squared distance from the center by twice the variance in that dimension. As in the case of hyperspheres, a classifier is active, if its activity $cl.ac$ lies above the threshold θ_m.

Again, we need to adjust covering, mutation and crossover. In covering, the center of the condition is again set to the current value, that is, $m = s$. The deviations σ_i are each chosen independently, uniformly randomly between zero and r_0 (zero excluded). Mutation is done as in the case of hyperspheres only that for each dimension each separate σ_i is considered. If any σ_i is larger than the deviation necessary to contain the whole problem dimension, it is set to that value, that is:

$$\forall i : \sigma_i \leq \frac{u_i^* - l_i^*}{\sqrt{-2\log\theta_m}}. \tag{9.10}$$

The crossover operator is uniform crossover over all $2n$ parameters of a condition part. Finally, a classifier is regarded as more general if it completely contains the other classifier in all n dimensions considered separately.

General Hyperellipsoids

Still, in all condition structures above, the condition structure is axis-dependent. It is possible to loosen the axis-dependency further using general hyperellipsoidal conditions. The condition part does not only now consist of a center and the length of each axis, but additionally considers interactions between the axis using a full matrix. The new condition part is defined as

$$C = (m_1, m_2, ..., m_n, \sigma_{1,1}, \sigma_{1,2}..., \sigma_{n,n-1}\sigma_{n,n}). \tag{9.11}$$

The condition is represented by a center point and a matrix. The matrix does not need to be restricted in any way—fully handing over the responsibility of evolving proper matrices to the evolutionary component. The new activity of a classifier is defined as

$$cl.ac = \exp\left(-\frac{\sum_{i=1}^{n}(\sum_{j=1}^{n}(s_j - m_j)\sigma_{ij})^2}{2}\right), \tag{9.12}$$

effectively multiplying the difference vector $s - m$ with the matrix, which effectively specifies the inverse covariance matrix. The matrix multiplication allows any rotations in the n-dimensional space as well as any possible squeezing and stretching. The hyperellipsoids of the previous section are a special case of these general hyperellipsoids. When only the matrix diagonal contains non-zero values, the condition is equivalent to a condition in the previous section with equal center and inverse deviation values ($\sigma_i = 1/\sigma_{ii}$). Classifier activity is again controlled by threshold θ_m.

In covering, the center of the hyperellipsoid is set to the current value. Only the diagonal entries in the covariance matrix are initialized to the squared inverse of an uniformly randomly chosen number between zero and one. All other matrix entries are set to zero. In this way, covering creates condition parts that are on average identical to the ones created by the covering mechanism in the previous (axis-dependent) hyperellipsoids.

Mutation is similarly adjusted in that each matrix entry is mutated separately maximally decreasing (increasing) the value by 50%. If the value was set to zero, it is initialized to a randomly chosen value as in covering for the diagonal matrix entries. The values of the matrix entries are unrestricted. Uniform crossover is applied to all $n + n^2$ condition part values. It is hard to determine exactly if a condition is more general than another condition. A useful approximation is that a condition is more general if its interval contains the interval of the other classifier plus the distance to the other classifier in the direction of that other classifier.

9.4.2 Performance Comparisons

The XCSR system has been evaluated on several one dimensional functions as well as on a six dimensional roots mean squared function (Wilson, 2002). The results clearly showed the advantage of the linear approximation instead of the constant one. However, the high dimensional function had only a rather weak curvature and thus was limited in the amount of linearity.

This section evaluates the different condition structures on several challenging test problems, each with its own kind of challenging non-linearity. Note that we restrict the number of actions to one in this section, degrading the action part in XCS to a dummy value. However, XCSR can be certainly be applied with linear predictions and multiple actions as has been investigated in Wilson (2004). The following three two-dimensional functions are tested:

$$f_1(x, y) = ((x * 3)\%3)/3 + ((y * 3)\%3)/3 \tag{9.13}$$
$$f_2(x, y) = (((x + y) * 2)\%4)/6 \tag{9.14}$$
$$f_3(x, y) = \sin(2\pi(x + y)), \tag{9.15}$$

where the % is the modulo operator. All the functions are defined for values between zero and one in both dimensions. Figure 9.2 shows the three functions. Function f_1 is an axis-parallel step function. It is expected to be learned quite effectively by XCSR with hyperrectangular conditions since the steps are rectangular so that it is possible to represent each step with one classifier. Function f_2 is an axis-diagonal step function. XCS faces oblique boundaries so that the approximation is expected to be significantly harder. Finally, function

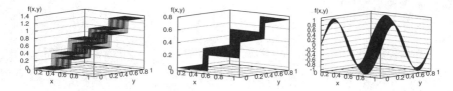

Fig. 9.2. The axis-parallel step function f_1, the axis-diagonal step function f_2 and the sinusoidal function f_3.

f_3 is an axis-diagonal, continuous sinusoidal function. Additional challenges due to the curvature are expected.

Performance of XCSR

Predictive performance of XCSR with hyperrectangles is shown in Figure 9.4[1]. The results confirm our expectations. Function f_1 is learned best resulting in a low error of about .007—significantly below the error threshold ε_0. The axis-diagonal step function is initially slightly easier to approximate due to its smaller y-value range. While learning proceeds, the population size rises to a higher level and the prediction just reaches .01. Due to the obliqueness of the function, the hyperrectangular conditions make it harder to evolve an effective space partitioning. However, the oblique structure is still easier to represent than the sinusoidal function. In f_3, XCSR only reaches an error level of .011. Learning takes longer and the consequent population size is much higher.

Function approximation performance is shown in Figure 9.3a, b, c (resolution 50 × 50). In the parallel-step function, XCS approximates the steps accurately. No severe under- or overestimations occur except at the boundaries. The condition structure is the main obstacle in functions f_2 and f_3 since the steps in the diagonal-step function are oblique and the sinusoidal function additionally has a strong curvature. Several approaches are imaginable to improve performance of XCSR. If the typicality of the function was known, then performance could certainly be improved. For example, in the case of the diagonal step function, an enhancement with diagonal boundaries should strongly improve performance. It should also be helpful in the sinusoidal function. However, it cannot be expected that such boundary types are known beforehand.

---

[1] Experiments are averaged over 20 experiments. If not stated differently, parameters were set as follows: $N = 6400$, $\beta = \eta = 0.5$, $\alpha = 1$, $\varepsilon_0 = .01$, $\nu = 5$, $\theta_{GA} = 50$, $\chi = 1.0$, $\mu = 0.05$, $r_0 = 1$, $\theta_{del} = 20$, $\delta = 0.1$, $\theta_{sub} = 20$. GA subsumption was applied. Uniform crossover was applied. Note that we start with a fairly general population due to the high covering value r_0. We also apply a rather high learning rate that showed to slightly speed-up performance due to faster evolving approximations. If the problems were noisy problems it would be necessary to decrease the learning rate.

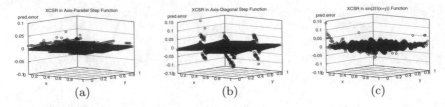

Fig. 9.3. Average approximation error in various function approximate tasks. (a) The boundaries of the axis-parallel function are well-approximated by XCSR. (b) In the case of oblique steps, XCSR misses some of the boundary points. (c) In the sinusoidal function, XCSR under- or overestimates. Corners and areas with strong curvature are affected most severely.

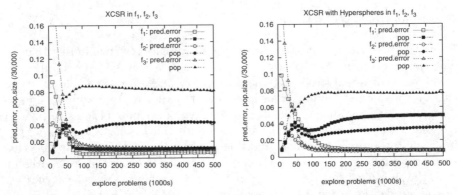

Fig. 9.4. XCSR with hyperrectangles performs well in the axis-parallel function f_1. Performance is worse in the axis-diagonal function. In the sinusoidal function, XCSR hardly reaches the target level of 1% error.

Fig. 9.5. While the axis-parallel step function is harder to learn for XCSR with hyperspheres, the axis-diagonal as well as the sinusoidal functions are approximated more easily.

A more general approach lies in a boundary approximation by the means of circular or ellipsoidal structures as introduced above. Although these structures cannot be expected to yield perfect approximations, the piecewise linear approximations should suitably overlap, representing a smoother approximation surface. Disruptive effects in the corners of hyperrectangles are expected to be prevented.

Hyperspheres

Performance of XCSR with hyperspheres is shown in Figure 9.5. In the axis-parallel step function, XCSR with hyperspheres reaches a performance level comparable to that of hyperrectangular conditions. Since hyperspheres are not well-suited to approximate axis-parallel boundaries, the condition structure complicates the task so that more learning time is required and more classifiers are needed for an equally accurate approximation. The roles are reversed in the axis-diagonal step function: With hyperspheres, XCSR needs less classifiers to reach a better performance than with hyperrectangles. Also learning is faster. Similar performance differences are observable in the sinusoidal function. The final population size is again smaller and learning proceeds faster. Moreover, the final approximation is more accurate when hyperspheres are applied. In both cases, the unnatural space partition of an oblique function via hyperrectangles (1) requires more specialized final classifiers and (2) causes unsuitable approximations particularly in the corners of the rectangles.

Figures 9.6a, b, c show the three-dimensional error surfaces when spheroidal conditions are used in XCSR. XCSR does not approximate the axis-parallel steps completely accurately. However, the surfaces inside the steps are approximated more accurately so that the performance average over the whole problem reaches a similar level. The axis-diagonal step function as well as the sinusoidal function are both better approximated when hyperspherical conditions are used. In the case of the axis-diagonal step function, the spherical condition structure is slightly recognizable since the errors at the steps are slightly bent, indicating that a circle tends to overlap with the boundary. The performance error remains smaller since the overlap is less severe than in the case of hyperrectangles.

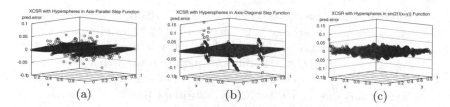

Fig. 9.6. (a) The axis-parallel steps are slightly harder to approximate with hyperspherical conditions. (b) In the case of oblique steps, hyperspherical conditions allow a better approximation. (c) The sinusoidal function can be approximated better with hyperspherical conditions.

Besides the performance curves it is interesting to investigate how XCSR evolves the generality of its conditions. We expect that XCSR evolves more specialized conditions in regions in which the curvature of the function is highest since in those regions linear predictions are hardest to fit. Figure 9.7 confirms this suspicion in the three functions. In the step functions, deviations differ between the steps. At the steps, more specialized classifiers emerge. In the sinusoidal function, the standard deviation (or generality) of the condition parts of the matching classifiers directly depends on the curvature of the underlying function. The outlayers are located at the top and bottom corner

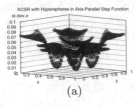 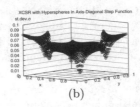 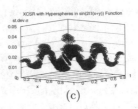

(a)   (b)   (c)

Fig. 9.7. In the axis-parallel (a) and axis-diagonal (b) step functions, the standard deviations of classifier conditions clearly exhibit the functional properties. More specialized conditions evolve for the interior steps because of the bounds in both dimensions. In the sinusoidal function, the deviations correlate with the curvature of the function (c).

of the space (that is, around 1,0 and 0,1). Since the occurrence in this region is small, the classifiers tend to cover larger regions.

As mentioned above, spheroidal conditions assume that nonlinearities change equally strongly in all dimensions. If the complexity of the target function differs in different dimensions, XCSR with the simple hyperspheres cannot be expected to learn very well since hyperspheres are equally spaced in all dimensions. Thus, we investigate the following function:

$$f_4(x_1, x_2, x_3, x_4) = x_2 + x_4 + \sin(2\pi x_1) + \sin(2\pi x_3) \qquad (9.16)$$

This four-dimensional function is easily linearly approximated in the x_2 and x_4 dimensions. Dimensions one and three are much harder to approximate. Figure 9.10 shows performance of XCSR with hyperrectangles and hyperspheres. Clearly, the hyperrectangular encoding outperforms the hyperspheroidal encoding. To represent the problem effectively, the space should be partitioned only in dimensions one and three. The additional partitioning in the other two dimensions forced by the spheroidal encoding hinders the evolution of more effective space partitions. Hyperellipsoidal conditions do not make such a restrictive assumption.

Hyperellipsoids

Performance in the test functions f_1, f_2, f_3 (Figure 9.8) is similar to that of the hyperspherical ones. The population sizes converge to a higher level, though, due to the additional variability in the conditions. Also the three dimensional error surfaces do not differ significantly (not shown).

Performance in the crucial f_4 function, however, is strongly improved. Figure 9.10 shows that hyperellipsoids enable XCSR to reach a similar performance level to the representation with hyperrectangles. Note that XCSR with hyperellipsoids however does not beat the performance of XCSR with hyperrectangles in f_4. This can be explained by the independence of the four

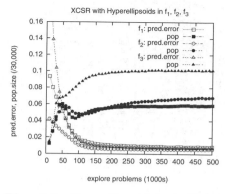

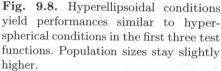

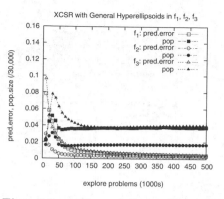

Fig. 9.8. Hyperellipsoidal conditions yield performances similar to hyperspherical conditions in the first three test functions. Population sizes stay slightly higher.

Fig. 9.9. General hyperellipsoidal conditions further improve performance compared to (restricted) hyperellipsoidal conditions in the first three test functions. Population sizes are also smaller.

dimensions: the hyperrectangular encoding basically approximates the underlying function, evolving linear approximations for each dimension. The hyperrectangular form causes the independence assumption in the four dimensions. However, if there are dependences like in the diagonal-step function or the sinusoidal function, the assumption is violated and the approximation suffers. Spherical or the more general ellipsoidal conditions alleviate this independence assumption.

Nonetheless, the hyperspherical conditions also obey the dimensionality in that the axes of the hyperspheres coincide with the dimensional axes. The independence assumption still exists—albeit in a less severe form. General hyperellipsoidal conditions abolish this assumption completely.

General Hyperellipsoids

Performance of XCSR with general hyperellipsoidal conditions in functions f_1, f_2, f_3 is shown in Figure 9.9. Function approximation is improved in all three cases and population sizes also reach a smaller level. XCSR is able to exploit the additional freedom in the hyperellipsoidal structures, turning and tweaking them to the most effective subspace approximations. In the f_4 problem, XCSR's performance with general hyperellipsoids stays slightly above the performance with restricted hyperellipsoids, indicating that the abolished independence assumption between the axis makes the problem harder (Figure 9.10).

The power of general hyperellipsoidal conditions becomes most obvious in the three dimensional sinusoidal function:

$$f_5(x_1, x_2, x_3) = \sin(2\pi(x_1 + x_2 + x_3)). \tag{9.17}$$

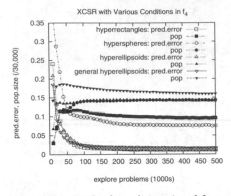
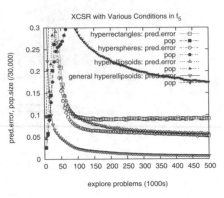

Fig. 9.10. In the four-dimensional function f_4, spheroidal conditions are outperformed by hyperrectangular ones. Hyperellipsoidal conditions alleviate the restriction.

Fig. 9.11. In the three-dimensional sinusoidal function, only the general hyperellipsoidal conditions enable accurate and effective function approximation with XCSR.

This function is not only more difficult in the additional dimension in comparison with the sinusoidal function f_3, but it also contains three full sinuses instead of two. Figure 9.11 shows that only XCSR with general hyperellipsoidal conditions reaches an error of less than .01. The decrease in population size confirms that XCSR finds a suitable ellipsoidal shape for the condition part. The shape spreads throughout the population, which results in the observable population size decrease. With hyperspheres or (restricted) hyperellipsoids, performance is worse than with general hyperellipsoids but clearly beats the hyperrectangular approximation.

9.5 Summary and Conclusions

This chapter has shown that XCS can be successfully applied to various real-valued problems. In datamining applications, XCS showed competitive performance in comparison to various other machine learning algorithms. The comparison included the decision tree learner C4.5, the rule extraction mechanism PART, the Naive Bayes classifier, as well as the support vector machine algorithm SMO. None of the algorithms clearly outperformed XCS, nor were any of the algorithms clearly outperformed by XCS (except majority and 1-R). The comparison with simple majority classification and the simple 1-R algorithm clearly confirmed successful learning in the XCS classifier system.

Additionally, we showed that the facetwise LCS theory established in the previous chapters can be extended to real-valued features, nominal features, as well as a combination of different types of features. In order to extend the theory, it is necessary to adjust the specificity measure to the condition representation at hand. In a real-valued representation, volume can be equated with generality, which corresponds to one minus specificity. With this redefinition at hand, the other parts of the theory can be adapted fairly easily.

Additionally, we showed that XCS's predictions can be enhanced to piecewise linear predictions, and conditions can basically be any kernel-based basis structure. The more suitable the condition structure to partition the non-linear problem space into piecewise linear approximations, the better XCS can learn the partition. Covering, mutation, and crossover need to be adjusted to make the evolutionary search most effective. Most recently, Lanzi, Loiacono, Wilson, and Goldberg (2005) has shown that non-linear—such as polynomial—predictions can also be used in XCS. Thus, the evolutionary mechanism searches for space partitions that enable maximally accurate predictions dependent on the prediction structure and the problem at hand. Thus, the more suitable condition structure and prediction structure are to the problem at hand, the faster XCS will evolve the problem solution.

It should be emphasized once more that XCS is not really an algorithm that is designed to do classification nor direct function approximation. In fact, all the XCS results in this chapter were obtained in a reinforcement learning framework. In contrast to the other learners that all apply statistical analyses on the whole train-dataset, XCS learns a competitive solution online receiving and classifying one training instance at a time. Feedback is provided in terms of instant reinforcement. XCS is designed to also solve problems in which reinforcement is not immediate, may vary over time, and may be delayed in time. The next chapter investigates these capabilities in various multistep reinforcement learning problems.

XCS in Reinforcement Learning Problems

The performance investigations of XCS in various classification problems including diverse Boolean function problems as well as datamining problems showed that the facetwise LCS theory can predict XCS's behavior accurately. We confirmed the importance of appropriate selection pressure, population sizing and specificity, as well as the dependence on problem properties such as the minimal order complexity, infrequent niche occurrences, or overlapping solution representations. The datamining applications confirmed XCS's machine learning competitive learning behavior in terms of accuracy and solution generality.

Thus, it is now time to face the last part of our facetwise LCS theory approach: the additional challenges in multistep problems. In particular, (1) effective behavior needs to be assured, (2) problem sampling issues may need to be reconsidered, and (3) reward needs to be propagated most accurately.

As shown in our XCS introduction, XCS learns an approximation of the underlying Q-value function using Q-learning. Thus, to enable learning of the Q-value function, behavior needs to assure that all state transitions are experienced infinitely often in the long run. However, since Q-learning is *policy independent*, the behavioral policy does not need to depend on the current Q-value estimates enabling other behavioral biases. Since learning was shown to be most successful if all problem niches are encountered equally frequently, a behavioral bias that strives to experience all environmental states equally often can be expected to be beneficial. However, in this chapter, we are interested in *whether* XCS can learn optimal behavior and an optimal Q-function approximation, rather than if learning can be further optimized by improving behavioral issues.

Most recently, Kovacs and Yang (2004) confirmed that more uniform problem sampling can strongly improve performance in multistep problems. Introducing an additional state-transition memory to XCS that is used to re-sample previously seen transitions, uniform state-transition sampling is approximated increasing XCS's learning speed. Several sampling issues are also visible in our experimental investigations in several maze and blocks-world scenarios. In effect, population sizes need to be adjusted to prevent niche loss and ensure reproductive opportunities, especially early in the run.

Most importantly though, this chapter considers the importance of an accurate reward propagation. In essence, we implement gradient-based update techniques in the XCS classifier system derived from results in the function approximation literature in RL (Sutton & Barto, 1998). These mostly neural-based function approximation techniques have been shown to be highly unstable if direct gradient methods are used to implement Q-learning (Baird, 1995). Accordingly, residual gradient methods have been developed to improve robustness (Baird, 1999).

The aim of this chapter is to explore XCS's performance in multistep problems. We investigate the possibility of applying gradient-based update methods in the XCS system, improving learning reliability and accuracy. We show how LCSs are related to neural function approximation methods and use similar gradient-based methods to improve stability and robustness. In particular, we apply XCS to large multistep problems using gradient-based temporal difference update methods, reaching higher robustness and stability. The system also becomes more independent from learning parameter settings.

The remainder of this chapter first gives an overview of (residual) gradient approaches in the RL-based function approximation literature. Next, we incorporate the function approximation techniques in the XCS classifier system. The subsequent experimental study confirms the expected improved learning stability and reliability in various multistep problems including noisy problems and problems with up to ninety additional randomly-changing irrelevant attributes. Summary and conclusions reconsider the relation of the XCS system to tabular-based RL and to function-approximation-based RL.

10.1 Q-learning and Generalization

Tabular Q-learning is simple and easy to implement but it is infeasible for problems of interest because the size of the Q-table (which is $|S| \times |A|$) grows exponentially in the number of problem features (attributes). This is a major drawback in real applications since the bigger the Q-table the more experiences required to converge to a good estimate of the optimal Q-table (Q^*) and the more memory required to store the table (Sutton & Barto, 1998).

To cope with the complexity of the tabular representation, the agent must be able to *generalize* over its experiences. That is, it needs to learn a good approximation of the optimal Q-table from a limited number of experiences using a limited amount of memory. In RL, generalization is usually implemented by function approximation techniques: The action value function $Q(\cdot, \cdot)$ is seen as a function that maps state-action pairs into real numbers (i.e., the expected payoff); gradient descent techniques are used to build a good approximation of function $Q(\cdot, \cdot)$ from on-line experience.

Already in Chapter 2 we showed the Q-table for Maze 1 in Table 2.2. From another perspective, we can order the payoff levels to generate a surface of different payoff levels. In the Maze 1 problem, for example, there are 40 state-action pairs in total. Assuming a discount factor of $\gamma = 0.9$, 16 of them correspond to a payoff of 810, 21 correspond to a payoff of 900, and only three state-action pairs correspond to a payoff of 1000 (assuming a discount factor of $\gamma = 0.9$). The goal of function approximation techniques is to develop a good estimate of this payoff surface.

When applying gradient descent to approximate Q-learning, the goal is to minimize the error between the desired value that is estimated by "$r + \gamma \max_{a' \in A} Q(s', a')$" and the current payoff estimate given by $Q(s, a)$ as shown in Equation 2.4. For example, when applying function approximation techniques parameterized by a weight matrix W using gradient descent, each weight w changes by Δw at each time step t,

$$\Delta w = \beta(r + \gamma \max_{a' \in A} Q(s', a') - Q(s, a)) \frac{\partial Q(s, a)}{\partial w}, \qquad (10.1)$$

where β is the learning rate and γ is the discount factor. Following the notation used in the RL-approximation literature, $Q(\cdot, \cdot)$ represents the approximated action value function not the Q-table (Baird, 1995; Baird, 1999; Sutton & Barto, 1998). This weight update depends both (i) on the difference between the desired payoff value associated with the current state-action pair (that is, $r + \gamma \max_{a' \in A} Q(s', a')$) and the current payoff associated with the state-action pair $Q(s, a)$ and (ii) on the gradient component represented by the partial derivate of the current payoff value with respect to the weight. The gradient term estimates the contribution of the change in w to the payoff estimate $Q(s, a)$ associated with the current state-action pair. Function approximation techniques that update their weights according to the equation above are called *direct algorithms*.

While tabular reinforcement learning methods can be guaranteed to converge, function approximation methods based on direct algorithms, like the previous one, have been shown to be fast but unstable even for problems involving few generalizations (Baird, 1995; Baird, 1999). To improve the convergence of function approximation techniques for reinforcement learning applications, another class of techniques, namely *residual gradient algorithms*, have been developed (Baird, 1999). Residual gradient algorithms are slower but more stable than direct ones and, most importantly, they can be guaranteed to converge under adequate assumptions. Residual algorithms extend direct gradient descent approaches, which focuses on the minimization of the difference "$(r + \gamma \max_{a' \in A} Q(s', a') - Q(s, a))$," by adjusting the gradient of the current state with an estimate of the effect of the weight change on the successor state. The weight update Δw for Q-learning implemented with a *residual gradient* approach becomes the following:

$$\Delta w = \beta(r + \gamma \max_{a' \in A} Q(s', a') - Q(s, a)) \left[\frac{\partial Q(s, a)}{\partial w} - \gamma \frac{\partial}{\partial w} \left(\max_{a' \in A} Q(s', a') \right) \right],$$
(10.2)

where the partial derivative $\frac{\partial}{\partial w}(\max_{a' \in A} Q(s', a'))$ estimates the effect that the current modifications of the weight have on the value of the next state. Note that since this adjustment involves the next state, the discount factor γ must also be taken into account.

Direct approaches and *residual gradient* approaches are combined in *residual algorithms* by using a linear combination of the contributions given by the former approaches to provide a more robust weight update formulation. Let ΔW_d be the update matrix computed by a direct approach, and ΔW_{rg} be the update matrix computed by a residual gradient approach, then, the updated matrix for residual approach ΔW_r is computed as follows:

$$\Delta W_r = (1 - \phi)\Delta W_d + \phi \Delta W_{rg}.$$
(10.3)

By substituting the actual expressions of ΔW_d and ΔW_{rg} for Q-learning with *direct* and *residual gradient* approach (see Baird, 1999 for details) we obtain the following weight update for Q-learning approximated with a *residual approach*:

$$\Delta w = \beta(r + \gamma \max_{a' \in A} Q(s', a') - Q(s, a)) \left[\frac{\partial Q(s, a)}{\partial w} - \phi \gamma \frac{\partial}{\partial w} \left(\max_{a' \in A} Q(s', a') \right) \right].$$
(10.4)

Note that the residual version of Q-learning turns out to be basically an extension of *residual gradient* approach where the contribution of the next state ($\frac{\partial}{\partial w} \max_{a' \in A} Q(s', a')$) is weighted by the parameter ϕ. The weighting can be either adaptive, or it can be computed from the weight matrices W_d and W_{rg} (Baird, 1999).

With the reinforcement learning knowledge in mind, we now turn to the XCS classifier system investigating how XCS approximates the Q-function and how gradient-based updates can be incorporated into XCS.

10.2 Ensuring Accurate Reward Propagation: XCS with Gradient Descent

As shown before, XCS uses Q-learning techniques but can also be compared to a function approximation mechanism. In this section, we analyze the similarities between, tabular Q-learning, Q-learning with gradient descent, and XCS. We show how to fuse the two capabilities, adding gradient descent to XCS's parameter estimation mechanism.

10.2.1 Q-Value Estimations and Update Mechanisms

As explained above, while tabular Q-learning iteratively approximates Q-table estimates, a function approximation approach estimates the Q-table entries by the means of a weight matrix. Using the direct (gradient descent) approach, each weight w in the matrix W is modified by the quantity Δw determined by Equation 10.1. Hereby, the gradient component $\frac{\partial Q(s,a)}{\partial w}$ is used to guide the weight update.

As seen above, XCS exploits a modification of Q-learning applying direct and independent prediction updates to all classifiers in the to-be updated action set $[A]$. To see the relation directly, we restate the update rule here again for the multistep case (see also Equation 4.4):

$$R \leftarrow R + \beta(r + \gamma \max_{a \in [A]} P(a) - R). \tag{10.5}$$

We can note that while tabular Q-learning only updates one value at each learning iteration (updating $Q(s,a)$), XCS updates all classifiers in the action set $[A]$. In fact, each position in the Q-table is represented by the corresponding prediction array value (Equation 4.1). Comparing the weight update for gradient descent (Equation 10.1) and the update for classifier predictions (Equation 10.5) we note that in the latter no term plays the role of the gradient. Classifier prediction update for XCS was directly inspired by *tabular* Q-learning (Wilson, 1995) disregarding any gradient component.

10.2.2 Adding Gradient Descent to XCS

To improve the learning capabilities of XCS we add gradient descent to the equation for the classifier prediction update in XCS. As noted above, the value of a specific state-action pair is represented by the system prediction $P(\cdot)$, which is computed as a fitness weighted average of classifier predictions (Equation 4.1). In general, learning classifier systems consider the rules that are active and combine their predictions (their *strength*) to obtain an overall estimate of the reward that should be expected. In this perspective, the classifier predictions play the role of the weights in function approximation approaches. The gradient component for a particular classifier cl_k in the to-be-updated action set $[A]$ can be estimated by computing the partial derivate of $Q(s,a)$ with respect to the prediction p_k of classifier cl_k:

$$\frac{\partial Q(s,a)}{\partial w} = \frac{\partial}{\partial R_k} \left[\frac{\sum_{cl_j \in [A]} R_j F_j}{\sum_{cl_j \in [A]} F_j} \right] =$$

$$= \frac{1}{\sum_{cl_j \in [A]} F_j} \frac{\partial}{\partial R_k} \left[\sum_{cl_j \in [A]} R_j F_j \right] = \frac{F_k}{\sum_{cl_j \in [A]} F_j}. \tag{10.6}$$

Thus, for each classifier the gradient descent component corresponds to its relative contribution (measured by its current relative fitness) to the overall prediction estimate.

To include the gradient component in XCS's classifier prediction update mechanism, prediction R_k of each classifier $cl_k \in [A]$ is now updated using

$$R_k \leftarrow R_k + \beta(r + \gamma \max_{a \in A} P(a) - R_k) \frac{F_k}{\sum_{cl_j \in [A]} F_j}. \tag{10.7}$$

All other classifier parameters are updated as usual (see Chapter 4).

Due to the contribution-weighted gradient-based update, the estimate of the payoff surface, that is, the approximation of the optimal action-value function Q^*, becomes more reliable. As a side effect, the evolutionary component of XCS can work more effectively since the classifier parameter estimates are more accurate. The next section validates this supposition.

10.3 Experimental Validation

We now compare XCS's performance without and with gradient descent on several typically used maze problems as well as a blocks world problem. The results confirm that the incorporation of the gradient-based update mechanism in XCS allows the system to reliably learn a large variety of hard maze problems. In addition, we show that XCS is able to ignore irrelevant attributes requiring only a linear increase in population size. Finally, we apply XCS to the maze problems with additional noise in the actions.

The results confirm that XCS is a reliable reinforcement learning algorithm that is able to propagate gradient-based reinforcement backward and evolve a very general, accurate representation of the underlying MDP problem.

Parameters were set slightly different from the ones in the previous chapter, slightly increasing the population size and decreasing the don't care probability to prevent niche loss and ensure the capability of handling large problem spaces, respectively. If not stated differently parameters were set to $N = 3000$, $\beta = 0.2$, $\alpha = 1$, $\varepsilon_0 = 1$, $\nu = 5$, $\theta_{GA} = 25$, $\chi = 1.0$, $\mu = 0.01$, $\gamma = 0.9$, $\theta_{del} = 20$, $\delta = 0.1$, $\theta_{sub} = 20$, and $P_\# = 0.8$ throughout the experiments. Uniform crossover was applied. Tournament selection is applied throughout this section with a tournament size proportion of $\tau = 0.4$ of the action set size. Results are averaged over fifty experiments in the usual case but over twenty experiments in the case of additional sixty or ninety random attributes.

The experiments alternated between learning (exploration) and testing (exploitation) trials. During a learning trial, actions were executed at random whereas during testing, the best action was chosen deterministically according to the prediction array values. Additionally, during testing the evolutionary algorithm was not triggered but reinforcement updates were applied (as used in the literature (Wilson, 1995; Lanzi, 1999a)). If a trial did not reach the goal position after fifty steps, the trial is counted as a fifty step trial and the next trial starts. Lanzi (1999a) showed that such a reset method can improve learning performance since it assures that the search space is explored more equally. For the sake of comparison between gradient and non-gradient-based updates in XCS, however, the advantage applies to both systems and is therefore irrelevant. Additionally, we assure that a trial undergoes a maximum

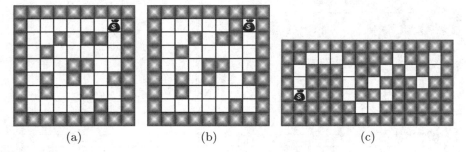

Fig. 10.1. Maze5 (a), Maze6 (b), and Woods14 (c) all pose different challenges to the XCS learning algorithms.

number of fifty learning steps so that the actual number of learning steps is bound by the fifty steps.

10.3.1 Multistep Maze Problems

We apply XCS to three different maze problems previously studied with the XCS classifier system (Lanzi, 1999a) and other LCS systems (Butz, 2002). The investigated Maze problems are shown in Figure 10.1. Maze5 and Maze6 are very similar in nature. However, the Maze6 environment poses additional hard challenges since it is much less likely to reach the food position during a random walk. Woods14 is challenging due to the long backpropagation chain— in total eighteen steps that need to be back-propagated.

The maze problems are coded similar to the Maze 1 and Maze 2 problems introduced in Chapter 2. A state in the environment is perceived as a binary feature vector of length sixteen, coding each of the eight neighboring positions by two bits starting north and coding clockwise. An obstacle is coded by 01, food (or money) is coded by 11, and finally an empty position is coded by 00. Code 10 has no meaning. A payoff of 1000 is provided once the food position is reached and the agent is reset to a randomly selected empty position in the maze.

Figure 10.2 compares performance in Maze5 (a,b) and Maze6 (c,d) without and with the gradient-based update technique. Clearly, XCSTS with gradient-based update learning is able to learn the optimal path to the food from all positions in the mazes whereas XCSTS without gradient-based updates fails to learn an optimal path and stays basically at random performance. Investigations in the evolved payoff landscape showed that XCS with gradient strongly tends to evolve an overgeneral representation that result in an inaccurate high-variance reward prediction. XCSTS with gradient, on the other hand, evolves the underlying payoff landscape accurately. Also with ten additional random attributes, which are set uniformly randomly at each step, XCSTS without gradient does not reach any better performance. XCSTS with

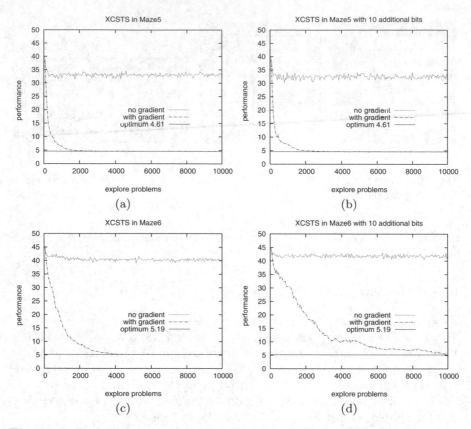

Fig. 10.2. Performance comparisons in Maze5 and Maze6 confirm the superiority of the gradient approach.

gradient again reaches optimal performance although slightly delayed due to the small population size of $N = 3000$.

Increasing the population sizes, XCS is even able to learn an optimal policy if 30 ($N = 5000$), 60 (N=7500), or 90 (N=10,000) bits are added to the perception (Figure 10.3a,c). These bits are changing at random and are consequently highly disruptive in the learning process. Nonetheless, XCS is able to identify the relevant bits quickly and ignore the irrelevant attributes. The monitored population sizes (Figure 10.3a,c) reach a very high value initially but then quickly degrade to a low level. Hereby, as expected, the minimum population size is not reached since mutation continuously introduces specializations in the irrelevant bits and thus additional classifiers in the population. Subsumption can decrease the population size only slowly since the reward propagation continuous to be noisy so that classifier accuracy ($\varepsilon < \varepsilon_0$) is hard to reach.

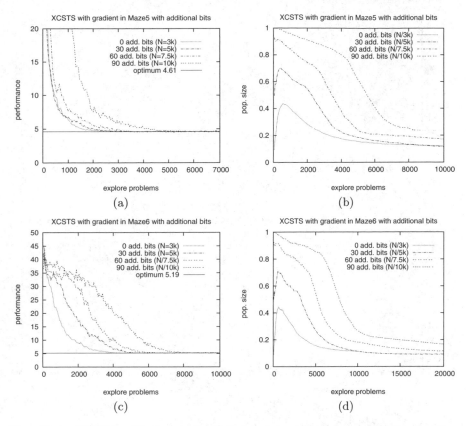

Fig. 10.3. XCSTS with the gradient approach is able to learn an optimal policy even when adding 90 randomly changing bits. The population sizes show that although the additional random attributes result in a very large initial population size, population size drops off quickly, indicating the evolving focus on relevant attributes.

Similar observations can be made in Woods14 as shown in Figure 10.4. In Woods14 it seems particularly important to have a larger population size available. Due to the long chain of necessary back-propagations and the infrequent occurrence of distant states in the environment, problem sampling issues need to be reconsidered. In particular, due to the infrequent sampling of problem subspaces, niche support as well as reproductive opportunities can only be assured if the population is large enough. Due to the disruption when the population size is set too small, learning the optimal behavioral policy is delayed (Figure 10.4 a,b) in contrast to the sufficient size of $N = 5000$ (Figure 10.4 c).

This observation is also confirmed when analyzing the single Woods14 runs: Out of the fifty runs, four had not converged after 5000 exploration problems with a population size of $N = 3000$. Due to the skewed problem

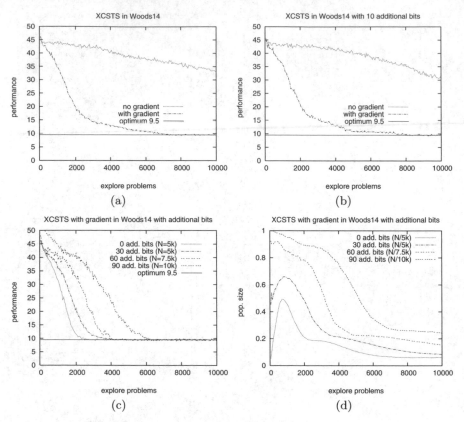

Fig. 10.4. Also in Woods14, the gradient-based update approach is highly beneficial. Also a higher population size yields better performance (a vs. c). Additional random attributes show the resulting strong learning potential of the XCSTS system with gradient-based updates.

sampling, reproductive opportunities cannot be assured when the population size is set too low, disrupting learning in some runs. The higher population size assures reproductive opportunities and thus assures fast learning in all runs. Thus, as expected by the facetwise LCS theory, extra care needs to be taken in multistep problems due to expectable highly skewed problem sampling.

Besides the performance curves, we are also interested in the actual approximation of the underlying Q-value function. According to the theory above, XCS should learn an exact approximation of the Q-value function. Figure 10.5 shows the maximum Q-value estimate in each state of XCSTS (a,c) and XCSTS (b,d) with gradient in Maze5 (a,b) and Woods14 (c,d) including the standard deviation of the estimations over fifty runs after 10,000 learning trials. Clearly, XCSTS with gradient is able to evolve an accurate representation of the Q-function. XCSTS without gradient, though, does not evolve accurate

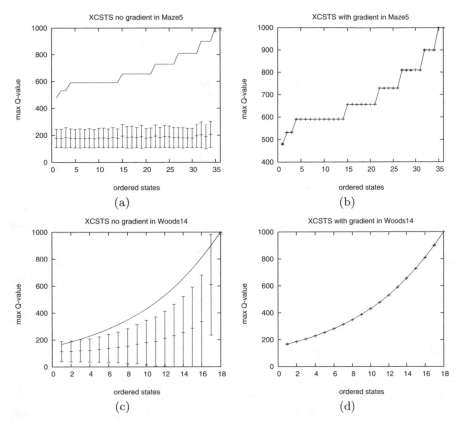

Fig. 10.5. The actual Q-value estimates confirm that XCS with gradient reliably evolves accurate Q-value approximations.

representations but stays overgeneral. In Woods14, the matter appears slightly better as also indicated by the performance curves in Figure 10.4b. However, the standard deviation is very high indicating high noise in the estimation values.

Finally, we investigated if XCSTS with gradient is also able to handle noisy problems. Consequently, we added noise to the actions of the system so that actions do not necessarily lead to the intended position but with a low probability, the actions leads to one of the neighboring positions. That is, with low probability an action to the north may lead to the position in the northwest or northeast. In our experiments, we set the probability of an action slip to 0.2. Results are shown in Figure 10.6 comparing performance in the case of a slippery surface to that in the deterministic problem. The results show that XCSTS with gradient is able to learn an accurate reward distribution in a noisy environment, allowing it to evolve a stable, near-optimal performance. If additionally adding thirty randomly changing attributes, convergence is

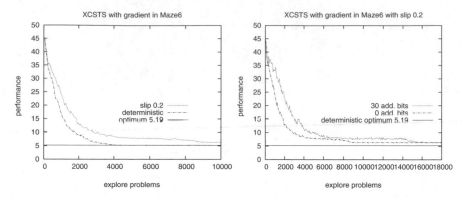

Fig. 10.6. Adding non-determinism to the actions delays the evolution of an optimal performance. Nonetheless, even with thirty additional random bits (right-hand side) XCS converges to a stable, optimal policy.

delayed further but after about 15000 learning trials also the last run of the twenty runs converges to the performance level of the one without additional random attributes.

10.3.2 Blocks-World Problems

Blocks worlds problems were investigated intensively over the last decades. Whitehead and Ballard (1991) studied temporal difference learning techniques in blocks worlds. The field of relational reinforcement learning studies logic-based relational representations. Kersting, Van Otterlo, and De Raedt (2004) show how to propagate reinforcement backward beginning from a formalized goal using logical inferences.

Our blocks world differs in that we code the blocks world perceptually instead of coding the explicit relations between the blocks. In fact, the relations need to be inferred from the online interaction with the blocks. This setup makes the problem more similar to an actual real-world scenario.

The blocks world is perceived from a global perspective, observing the distribution of blocks over the available stacks in the problem. Actions manipulate the blocks by gripping or releasing a block on a specific stack. Figure 10.7 shows several exemplar states and the corresponding perceptions in a blocks world with $s_{BW}=3$ stacks and $b_{BW} = 4$ blocks. The perception consists of b_{BW} positions for each stack, coding the presence of a block by 1 and the absence by 0. An additional bit codes if the gripper currently holds a block. Actions are possible of gripping or releasing a block on a specific stack. Thus, there are $2s_{BW}$ actions available in this blocks world. The goal is defined by transporting a certain number of blocks onto the first stack. For example, in the shown blocks world in Figure 10.7, the goal is to move all four blocks onto the first stack (right-hand side figure). Similar to the maze scenario,

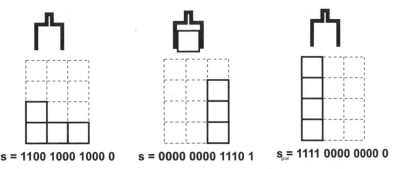

Fig. 10.7. The three exemplar states in the blocks world with $s_{BW}=3$ stacks and $b_{BW} = 4$ blocks illustrate the blocks world problem. The goal is defined as transporting a certain number of blocks onto the first stack.

XCS encounters a blocks world scenario receiving reward once the goal state is reached, triggering the start of the next trial.

Results in blocks world scenario are depicted in Figure 10.8. The figure on the left confirms once more the superiority of the gradient-based update. The goal to put three blocks onto the first stack is still rather easy to accomplish and also the setting without gradient evolves an optimal policy, albeit more slowly. Due to the low required height of the target stack, reaching the goal by chance occurs frequently, which triggers the provision of external reinforcement frequently, which again provides a much stronger reinforcement signal. Increasing the required block height of the goal stack by one makes it harder to evolve an optimal policy for XCS without gradient. Reaching the goal stack becomes infrequent, which makes it hard for XCS to learn a proper reward distribution. With the gradient-based update technique, the propagation of reward becomes much more stable and XCS is still able to learn the optimal policy.

Note that prioritized exploration algorithms might improve performance such as the Dyna-Q+ algorithm (Sutton, 1991), which prioritizes updates according to the delay since the most recent update, or the prioritized sweeping approach (Moore & Atkeson, 1993), which prioritizes exploration as well as internal RL updates according to the amount of change caused by the most resent update. The point of the gradient-based update technique, though, is to make XCS learning more robust. Prioritized exploration could be incorporated additionally to make learning even faster.

Figure 10.8 (right-hand side) shows performance of XCS with gradient in larger blocks world problems. XCS can solve problems as large as the 8x8 blocks world problem that codes a state by $l = 65$ bits and comprises 9876 distinct states and 16 possible actions. Thus, XCS is able to filter the input bits for relevancies with respect to the current problem and goal at hand, efficiently evolving a near-optimal behavioral policy. It appears that XCS does not exactly reach optimality, however, which is especially visible in the smaller

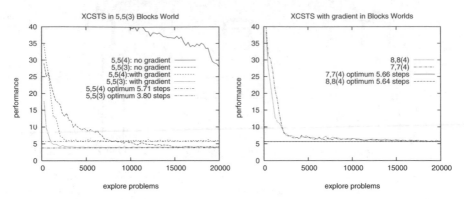

Fig. 10.8. Also in the blocks world problem, the gradient update techniques speeds-up and stabilizes learning of an optimal performance. Population sizes were set to $N = 5,000$ for the $s_{BW} = 7$, $b_{BW} = 7$ and to $N = 6000$ for the $s_{BW} = 8.b_{BW} = 8$ problem.

problems. The explanation is the states above the goal state, that is, when more than the actual required blocks are situated on the goal stack initially. This case occurs very infrequently so that the sampling of this subproblem space is very infrequent. Thus, a stable representation of the exceptional cases cannot evolve as predicted by the reproductive-opportunity bound and our facetwise LCS theory.. Thus, if this scenario occurs, XCS is not able to act optimally, which accounts for the slight in-optimality in the overall average performance.

10.4 Summary and Conclusions

This chapter has shown that XCS with tournament selection and gradient descent is able to solve a variety of multistep problems reliably and optimally. XCS showed to be noise robust as well as very powerful in ignoring additional irrelevant randomly changing attributes. Hereby, the population size needs to grow linearly in the additional number of features as predicted by our XCS theory.

Thus, we confirmed that the basic learning results from our facetwise theory approach carry over to multistep problems. However, multistep problems also showed to pose additional learning challenges as suggested in the facetwise LCS theory approach. Essentially, effective adaptive behavior needs to be ensured. Moreover, population sizes may need to be increased due to potentially skewed problem sampling. Most importantly, though, reward needs to be propagated and approximated most accurately.

The addition of the gradient-based reward-prediction update technique showed that the XCS learning architecture lies somewhat in between a tabular-based Q-learner and a neural-network based Q-learner. In tabular-based RL and in particular in Q-learning, each Q-value is represented by *one* entry in the Q-table. In neural-based reinforcement learning each Q-value is approximated by the activity in the *whole* neural network. XCS lies in between because it estimates Q-values by a *subset* of classifiers. The subsets distinguish different reward levels which explains why the residual gradient addition appears not necessary in XCS: The residual gradient distinguishes different, subsequent reward levels. In XCS, this distinction is evolved by the GA component.

In essence, XCS approaches function approximation by detecting subspaces whose function values can be effectively approximated by the chosen approximation mechanism. In the simplest case, the approximation is a constant as in the work herein but other methods, such as linear approximation, are under investigation (Wilson, 2001a). XCS's learning mechanism evolves such a distributed representation interactively. While the genetic component evolves problem space partitions that allow an accurate value estimation in each of the evolved subspaces, the RL-based approximation method estimates the value in each partition and rates the relative quality of each partition in terms of the accuracy of the value prediction. The gradient approach enabled a more reliable reward back-propagation and thus more reliable learning of an optimal behavioral policy.

In conclusion, this chapter showed that XCS is a valuable alternative to neural network based function approximation techniques. Especially when only few problem features (that is, sensory inputs) are relevant for the task at hand, XCS's computational requirements increase only linearly in the number of irrelevant features, hardly affecting learning speed. Thus, XCS serves well in RL-based problems, with many irrelevant and redundant problem features detecting and propagating the features relevant for accurate predictions reliably.

11

Facetwise LCS Design

The XCS classifier system is only one among many evolutionary rule-based learning systems. With the successful facetwise analysis approach carried through in XCS, it needs to be considered how the theory may carry over (1) to analyze other LCSs and related systems in the same way and (2) to apply the knowledge to create new LCS systems, targeted to a specific problem at hand.

So far, the facetwise analysis approach enabled several advancements in theory and understanding of the XCS classifier system. It was shown that different learning mechanisms cause different learning biases and can be substituted by other, potentially more effective biases. We now use the same facetwise analysis approach to investigate learning biases in other LCSs, relating and contrasting them to the biases uncovered in XCS.

The chapter explores each LCS learning facet and outlines the differences between various realizations of the same facet, such as differences between strength-based and accuracy-based fitness or tournament selection and proportionate selection. We discuss for which problems and objectives which learning mechanisms and consequent learning biases are most appropriate. Moreover, we highlight functional similarities of different LCS learning mechanisms and discuss the advantages and disadvantages of each mechanism.

11.1 Which Fitness for Which Solution?

Already in the introduction to simple LCSs we mentioned the importance of the correct fitness approach. Strength-based systems suffer from the problem of strong overgenerals (Kovacs, 2000). Although accuracy-based systems can be misled by unequally distributed variances in payoff, our experimental evaluations in Chapter 8 showed that XCS is very robust with respect to noise. Calculations in Kovacs (2003) showed that different noise values can also result in strong overgeneral classifiers in XCS. However, overgeneral classifiers can occur only in extremely unequally distributed noise settings, confirming XCS's robustness.

Nonetheless, accuracy-based fitness approaches might have other drawbacks with respect to the task at hand. Essentially, XCS's mechanism is designed to learn a reward map over the *complete* problem space as accurately as possible. Independent of the amount of reinforcement received, XCS strives to learn accurate reward predictions for all rewards. This is desirable when a complete reward map should be learned, which is necessary, for example, in RL problems, where XCS learns an approximation of the Q-value function. In other problem settings, reward might indicate the importance of a classifier representation rather than the correctness of applying that representation. In such scenarios, a different fitness approach seems more appropriate. In other words, if reward indicates the desired distribution of the problem representation, the learner should not learn an accurate representation of the distribution but rather it should represent the distribution itself in its classifiers. In such a scenario, more reproductive events and learning effort should occur in areas in which higher reinforcement is received.

Incidentally, the ZCS classifier system (Wilson, 1994) is essentially taking this road. ZCS is a strength-based system that applies *fitness sharing* to overcome the problem of strong overgenerals as shown in Bull and Hurst (2002). Additionally, ZCS applies the GA globally using proportionate selection based on fitness and deletion based on the inverse of fitness. Thus, ZCS reproduces classifiers with higher fitness, evolving a population of classifiers that converges to identical fitness values. Given the average fitness value $\overline{f}$ in the population, a problem niche in which a reward of r is encountered will be populated by $r/\overline{f}$ classifiers on average. Thus, ZCS should work very well in problems in which reward actually indicates representational importance. The recent successful application in the 37-multiplexer function confirms the potential of the system (Bull, 2003).

Thus, when comparing accuracy-based fitness with shared, strength-based fitness, it is clear that neither is always appropriate. If the task is to learn a complete and accurate reward map representation (including zero rewards), then accuracy-based fitness seems to be the right choice. If the task is to learn the highest reward cases accurately and the lower reward cases progressively less accurately, then the ZCS-framework should be more effective.

11.2 Parameter Estimation

In most parts of this book we assumed that parameter estimates accurately reflect the expected average parameter values in the long run. We additionally showed that in very noisy problems a lower learning rate β is necessary to assure dependable reward and reward prediction error estimates. In Chapter 10 we additionally showed the relation of the XCS parameter estimates to function-approximation techniques. The relations suggest that β should be a rather small value ($\beta \leq 0.2$) to filter noisy and unreliable input cases.

Somewhat surprisingly, Bull and Hurst (2002) showed that in the ZCS system larger β values might actually be advantageous. Both approaches are correct, though, and strongly depend on the dynamics in the population and the value the parameter intends to estimate.

In XCS, reward prediction and reward prediction error are classifier estimates that are determined *independent* of the rest of the classifier population. They are designed to approximate the expected average reward prediction and reward prediction error value as accurately as possible. Thus, the lower the learning rate β the more accurate the final estimate, particularly in noisy problems. However, the higher the learning rate β, the faster the initial estimate will be meaningful. This explains why it is very useful to apply the moyenne adaptative modifiée technique (Venturini, 1994) that applies initially large updates (effectively averaging the so-far encountered experiences) and then converges to the small β-based updates.

Fitness, on the other hand, reflects the relative accuracy of a classifier and thus does not apply the moyenne adaptative modifiée technique, but approximates its fitness value controlled by learning rate β from the beginning. Since an appropriate relative fitness estimation relies on an accurate error estimation, initial sufficiently low fitness updates are mandatory to prevent disruptive effects due to potential fitness over-estimation. Once a classifier is sufficiently experienced, though, it might be interesting to experiment with larger learning rates for the fitness measure to be able to adapt more rapidly to the current relative accuracy.

The successful application of very high learning rates β (up to one) in ZCS, which improved system performance in several problems (Bull & Hurst, 2002), can only be explained by an advantageous rapid adaptation to current population dynamics. Since classifier fitness in ZCS depends on the shared reward in a problem, the reward share directly reflects the actual coverage of the niche in question. The smaller the share, the more over-populated is the niche in question, as long as the provided reward indicates the proportional importance of each niche (as discussed above). It still remains to be investigated, though, in which problems a high learning rate β in ZCS might actually cause misleading estimates or niche loss. Learning disruption can be expected in problems with overlapping problem subsolutions or in highly noisy problems.

In XCS, a similarly high learning rate may actually be useful for updating the action set size estimate. Since the estimate solely depends on population dynamics, faster adjustments should lead to a more appropriate representation of the current population dynamics and should thus be advantageous. However, in overlapping problem domains, additional niching care needs to be applied.

As an alternative to varying learning rate β, Goldberg's variance-sensitive bidding approach for LCSs may be used (Goldberg, 1990). It is suggested that classifiers with high estimation variance should add additional variance to their reward prediction or fitness to prevent overestimations due to young or inaccurate classifiers. In this way, promising young classifiers may still get a chance to be considered for reproduction, making the evolutionary learning progress somewhat more noisy but stabilizing it once all classifiers converge to accurate values.

11.3 Generalization Mechanisms

Besides the search for a complete and accurate problem solution, it is important to evolve a maximally general solution. Following Occam's Razor, the evolutionary learning method needs to be biased towards favoring more general solutions. In XCS, we saw that there is an implicit generalization pressure, that is, the set pressure, which propagates *semantically more general* rules. There is also an explicit generality pressure, that is, subsumption pressure, which favors *syntactically more general* rules. Both pressures favor more general solutions as long as maximal accuracy is maintained.

Semantic generality refers to rule generality with respect to rule applicability, that is, the more often a rule applies in a given data set, the more general the rule is. Syntactic generality refers to generality with respect to the syntax of the rule, that is, the larger the size of the subspace the rule covers, the more general it is. We now compare several methods to stress the evolution of semantically general and syntactically general rules in an evolutionary learning system.

Semantic generalization can be realized, as done in XCS, by applying reproduction in match sets or action sets and deletion in the whole population. However, this approach may be misleading since the resulting reproduction bias may cause an over-representation of unimportant problem subspaces, as discussed above.

A similar approach for semantic generalization was proposed recently in the ZCS system (Bull, 2003). When taxing the generation of offspring in ZCS decreasing fitness, both parent and child have a very low reproduction probability (dependent on the amount of tax). The low probability of reproduction persists until the next time they take part in an action set, updating fitness and consequently restoring parts of the actual fitness dependent on the learning rate. Interestingly, the consequential generalization pressure is very similar to the one in XCS. As in XCS, the time until the next reproductive event depends on the frequency of niche occurrence. Additionally, in the ZCS system the taxed fitness makes the offspring more likely to be deleted, which might cause a problem in some scenarios with lots of problem niches. Consequently, two related fitness parameters could be introduced to ZCS: one for reproduction and one for deletion. The reproductive fitness measure should be taxed upon reproduction, whereas the fitness measure relevant for deletion should immediately reflect the actual fitness in the current set. Note that in XCS the initial fitness decrease in an offspring classifier has a somewhat similar effect, preventing the premature reproduction of under-evaluated classifiers. Since deletion is based on the action set size estimates and fitness is considered only in experienced classifiers, the taxed fitness has no immediate effect on deletion probability.

The taxation of offspring generation in ZCS as well as the set pressure in XCS cause an implicit semantic generalization pressure rather than an explicit pressure. Explicit pressures could be applied as well by installing an explicit measure of semantic generality. For example, the average time between rule activation could be recorded using again for example, temporal difference learning techniques. The resulting semantic generality measure could be used to bias offspring reproduction or deletion. This approach is implemented in the ACS2 system (Butz, 2002), which keeps track of the average time until activation of a classifier, and uses this measure to further bias deletion if accuracy is insignificantly different. In general, a two-stage tournament-based selection process can be implemented that first selects classifiers based on fitness, but takes semantic generality into account if fitness does not differ significantly.

The same approach could be applied with respect to syntactic generality. Instead of monitoring the average time until application, the syntactic generality of competing classifiers could be considered.

We saw that XCS uses subsumption deletion to prevent the generation of unnecessary over-specialized offspring once an accurate, more general classifier is found. However, subsumption is restricted to deterministic problem cases since it only applies once the classifier error estimate drops below $\varepsilon < \varepsilon_0$.

Another approach might bias the effect of mutation on specificity. Given that high accuracy is reached, mutation could mainly be changed to a generalizing mutation operator that may only cause generalizations in the offspring conditions. This approach was taken in ACS2 in conjunction with heuristic specialization in order to achieve a maximally general problem representation. However, the heuristic approach again relies on determinism in the environment and would need to be replaced with a noise robust mechanism. Generalizing mutation as well as subsumption rely on an appropriate setting of the error tolerance threshold ε_0. Since this threshold cannot be assumed to be available in general, only biased selection or deletion mechanisms appear applicable in the general case.

Regardless of the generalization method, it needs to be assured that the generality criterion does not compare totally unrelated classifiers. As mentioned before, different problem niches may require different syntactic generality levels. Moreover, different problem niches may be at different developmental stages, exhibiting potentially significantly different current generality. Thus, a generality competition should only be applied in subsets of similar classifiers, like the classifier subsets naturally defined by action sets.

11.4 Which Selection For Solution Growth?

The applied fitness approach guides selection for reproduction and deletion. There are two fundamentally different types of selection: *population-wide selection* and *niche selection*.

As we have seen, XCS applies niche selection for reproduction but population-wide selection for deletion. Apart from the implicit generalization effect analyzed in Chapter 5 and the effect on niching analyzed in Chapter 6, the combination also results in a general search bias. Since GA reproduction is directly dependent on niche occurrence, search effort is dependent on the frequency of niche occurrence. As shown in the niche support analysis in Chapter 6, the number of representatives of a niche is directly correlated with the frequency of niche occurrence so that more frequently occurring niches are searched more excessively and are represented by more classifiers. Parameter θ_{GA} is able to partially balance this bias to ensure the sustenance of less frequently occurring niches. However, the balancing effect is limited and decreases overall learning speed.

Thus, population-wide selection for reproduction may be advantageous if niche occurrence frequency is not correlated with the importance of learning a niche. However, even if the occurrence frequency reflects niche importance, different subspaces in the problem space may require different computational search effort to find an appropriate solution. The niching techniques in XCS assure that an underdeveloped niche is not lost. Additional search biases may be employed to enforce an equal learning progress in all niches.

The reliance on (potentially sparse) reinforcement feedback can cause additional learning problems. Since inaccurate feedback potentially results in highly noisy parameter estimates, particularly in young classifiers, reproduction needs to be balanced with evaluation, preventing niche loss as well as overproduction of under-evaluated, inaccurate classifiers. Restricting selection to the problem niche in which evaluation occurred naturally balances evaluation and reproduction. The XCS classifier system is doing just that.

Deletion usually serves additional purposes in the LCS realm. On the one hand, deletion should be biased to delete useless classifiers. On the other hand, deletion should be biased to delete classifiers in overpopulated niches. These two criteria are combined in the deletion criterion in XCS, biasing deletion on the action set size estimate as well as on the fitness estimate.

Besides the selection biases, it needs to be decided which selection technique to apply. The two direct competitors are proportionate selection and tournament selection (although alternatives exist, see e.g. Goldberg, 1989). As discussed in Chapter 2, proportionate selection is more appropriate for balancing problem niches possibly maintaining a large number of equally fit classifiers. Tournament selection strives to converge to the best individual and thus is best suited to propagate best solutions. In XCS, niche reproduction is designed to find the accurate, maximally general classifier in each problem niche. Thus, tournament selection works best. On the other hand, deletion is designed to assure niche sustenance and an equal distribution of action set sizes. Thus, proportionate selection is well-suited for deletion.

The ZCS system applies population-wide selection striving to maintain a larger number of equally important subsolutions via selection and deletion. The fitness sharing approach assures that equal fitness corresponds to equal importance. Thus, proportionate selection is the appropriate choice for reproduction and deletion in ZCS.

11.5 Niching for Solution Sustenance

Besides the problems of learning a problem solution, niching techniques need to assure solution sustenance. XCS's mechanism of choice is niche selection for reproduction in combination with population-wide selection for deletion. Since deletion is biased towards deleting larger niches and selection is biased towards reproducing in frequently occurring niches, problem subsolutions are sustained with a computational effort that is linear in the number of required subsolutions. We also saw that in overlapping problem solutions additional niching mechanisms might further stabilize performance.

ZCS niching is accomplished by fitness sharing in conjunction with proportionate selection. As shown in Harik (1994), niching based on sharing and proportionate selection assures niche sustenance in time exponential in the population size. Thus, it is very effective. However, in this case overlapping problem solutions can also interfere.

Thus, other techniques might be more appropriate, such as the diverse crowding techniques (De Jong, 1975; Mahfoud, 1992; Harik, 1994). For example, in the mentioned restricted tournament replacement (Harik, 1994; Pelikan, 2002), niche maintenance and niche support balance are further assured by restricting classifier replacement to the close neighborhood of the offspring. Thus, instead of selecting classifiers for deletion at random, classifiers with structural similarities may yield better deletion choices, leading to better subsolution maintenance.

In the most extreme case, niche-based deletion may be applied restricting the set of deletion candidates to the niche in which a classifier was reproduced. The ACS2 system applies this method, ensuring equal support of all problem niches (Butz, 2002). In the case of overlapping subsolutions, however, even niche-based deletion causes competition amongst overlapping classifiers. In this case, two mechanisms are imaginable: (1) Apply two-stage deletion that biases towards deleting the classifier that is represented more often; or (2) apply restricted tournament replacement techniques inside the action set to replace a most similar classifier. Both techniques can serve as niche stabilizers and should be investigated in problems in which overlapping problem solutions are inevitable (like in the count ones problem, see Appendix C).

11.6 Effective Evolutionary Search

Besides the resulting search biases caused by the fitness approach, reproduction, and deletion, the actual search in the neighborhood of currently promising solutions is guided by mutation and crossover operators. As discussed before, mutation is designed to search in the local, syntactic (genotypic) neighborhood of the current best solutions. Crossover, on the other hand, searches in the neighborhood defined by the two parental classifiers. Thus, while mutation in combination with selection searches for better solutions in the local neighborhood, simple crossover searches more globally, combining the information of two classifiers in the generated offspring rules.

Both mutation and crossover can be disruptive in that they may cause undesired search biases. Mutation may cause an unwanted specialization effect as observed in our evolutionary pressure analysis (Chapter 5), potentially causing the population to grow unnecessarily. Large mutation rates may additionally disrupt the neighborhood search, potentially resulting in completely random offspring generation. Dependent on the problem, both effects should be prevented. On the other hand, though, mutation is important to maintain diversity in the population, enabling the important search in the local neighborhood. Thus, a balanced mutation rate is important to maintain diversity and assure effective neighborhood search without causing too strong disruption.

Crossover may be disruptive as well, as already expressed in Holland's original schema theory (Holland, 1975) and as investigated in Chapter 7. Uniform crossover assumes attribute independence, consequently only recombining attribute value frequencies. One- and two-point crossover assume that neighboring attributes in the coded problem instances are correlated. Thus, in problems in which such a correlation is known, one- or two-point crossover should be applied. Additionally, dependent on if classifiers are selected for reproduction in problem niches or population wide, crossover can be expected to cause less or more disruptions, respectively, since classifiers in a problem niche can be expected to be more similar. On the other hand, crossing over two highly correlated classifiers may hardly cause any innovative search bias.

Competent crossover operators, such as the BB-based offspring generation or the Bayes net-based offspring generation, applied to XCS in Chapter 7, can overcome both potential drawbacks of simple crossover operators (i) preventing distribution and (ii) propagating innovation by combining previously successful substructures (Goldberg, 2002). Results of the combination with the XCS classifier system showed that the generated dependency structure learned from the patterns of the current best classifiers effectively biases the neighborhood search, replacing simple crossover operators and diminishing the importance of mutation. Interestingly, we used the *global* dependency knowledge to generate *local* offspring, effectively biasing local offspring generation towards the global neighborhood structure in the problem. Similar model-based offspring generation mechanisms can certainly be expected to work well in other LCS systems, given a problem structure in which local neighborhood search and simple crossover operators are not sufficient to evolve a complete problem solution.

11.7 Different Condition Structures

The above considerations apply to the binary realm with its ternary condition structure but are also relevant for any other type of condition. As we have seen in Chapter 9, condition structure in XCS can be extended to hyperrectangles but also hyperspheres or hyperellipsoids.

Each condition structure incurs certain biases due to assumptions in the underlying problem structure. Hyperrectangles assume the independence of each dimension. That is, non-linearities, which are alleviated by the incurred space partitions, are (implicitly) assumed to occur independently in each problem dimension. Hyperspheres assume that each dimension equally contributes to the prediction landscape in that each dimension contributes an equal amount of matching space. Axis-parallel hyperellipsoids assume possibly different contributions from each dimension, but the axis-parallel constraint prevents the formation of effective partitions of interdependent dimensions. Only general hyperellipsoids are able to do so.

Once the best condition structure was chosen, initialization, computational effort and effective search issues need to be considered. For example, population initialization and population size need to ensure that the necessary GA-induced search biases can apply. The covering bound specifies the interaction of initial condition specificity and population size. The schema and reproductive opportunity bound ensure the existence and growth of necessary solution structures. Solution sustenance may need to be reconsidered dependent on the condition structure, expected condition size (subspace occurrence), and condition overlap.

Next, mutation needs to be defined to search in the local neighborhood of the condition structure most effectively. Recombination should be designed such that disruptive effects are minimized and structure recombination is optimized. Competent BB-processing mechanisms may be extended to the chosen condition structure.

Figure 11.1 shows how spherical and ellipsoidal condition structures may be mutated and recombined. In the spherical case, the spheres can only be enlarged and shrunk so that mutation is limited to such effects. When conditions specify hyperellipsoids, those can be shrunk and enlarged in either dimension separately. A proportional increase or decrease in all dimensions may also be advantageous. In the general hyperellipsoidal case, rotation may be implemented as an additional mutation feature. Similarly, it may be advantageous to enable crossover that recombines information of size and orientation independently, as indicated in Figure 11.2. In this way, advantageous substructure

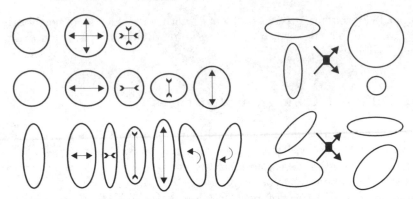

Fig. 11.1. While spheres only allow enlargement of shrinking, ellipsoids can be increased and decreased in each dimension independently. In the general case, mutation may also cause rotations.

Fig. 11.2. Crossover may be biased to combine different ellipsoidal structure information independently.

information may be effectively processed searching in the local neighborhoods but respecting global neighborhood knowledge.

11.8 Summary and Conclusions

This chapter gave further intuition about which mechanisms are responsible for which biases in LCSs. The facetwise approach enables the analysis of each learning aspect relevant for LCSs in separation. Various LCS mechanisms can cause similar or alternative learning biases considering a particular aspect of the facetwise theory. To decide which LCS system to apply, it is consequently very important to first consider the given problem structure and the expected problem properties. Dependent on the available feedback and the importance of problem instance occurrence, different fitness approaches and selection mechanisms are most appropriate. Similarly, condition structures should be tuned to the expected problem structure in that the chosen conditions should be able to represent necessary space partitions most effectively.

With the knowledge of this chapter in hand, it is hoped that the reader is ready to successfully design his own XCS or LCS system, tuned to the task at hand, adapted to the most suitable representation, exploiting expected problem properties. Accordingly, the next chapter outlines how these tools might be used to create cognitive LCSs as originally envisioned by John Holland (Holland, 1976; Holland & Reitman, 1978), who somewhat provocatively called the first LCS implementation the *cognitive system CS1*.

Towards Cognitive Learning Classifier Systems

This chapter outlines how the derived Facetwise LCS approach can carry over to the design of cognitive learning systems. We propose the integration of LCS-like search mechanisms into cognitive structures for structural growth and distributed, modular relevancy identification. The mechanisms may be integrated into multi-layered, hierarchical learning structures and may influence solution growth interdependently using only RL mechanisms and evolutionary problem solution structuring.

To regard a learning system as a *cognitive system*, the system should satisfy certain criteria. First, learning should be done online interacting with an outside environment. Second, learning may not be supervised in the sense that an error signal should not be provided directly from the outside environment. Only internal gradients and reinforcement can serve as learning signals. Third, the evolving knowledge should be distributed in a network structure to make it robust against failures of small computational entities (neurons). Fourth, learning should start from scratch with potentially many learning predispositions but without any explicitly coded knowledge.

This chapter proposes the design of cognitive learning architectures using a facetwise approach. Since the proposition of a complete facetwise approach for cognition is beyond the scope of this book, we focus on several cognitive aspects that appear to be most important. For each aspect, we outline how LCS-like learning mechanisms may serve as a valuable algorithm to design the desired learning structure adjusted to the cognitive task at hand.

In particular, we focus on the following four cognitive aspects:

1. **Predictive representations**: To be able to learn without any explicit teaching signal, predictions need to be learned that are adjusted using resulting outcomes.
2. **Incremental learning**: Online learning from scratch requires incremental learning growing and shaping the internal knowledge base.

3. **Hierarchical representations**: Our modular, hierarchical world composed of progressively more complex substructures needs to be reflected in the learning structure.
4. **Anticipatory**: Anticipatory mechanisms serve as the control mechanism controlling attention, action decision, and action execution. Violated predictions, expectations, and intentions serve as learning signals that guide and stabilize learning progress.

The next sections discuss these issues in turn, leading to the proposition of a multi-modular, hierarchical, anticipatory learning structure.

12.1 Predictive and Associative Representations

Predictive representations gained significant research interest over the recent years. The recent workshops on Predictive Representation of World Knowledge (Sutton & Singh, 2004) as well as the second workshop on Anticipatory Behavior in Adaptive Learning Systems (Butz, Sigaud, & Swarup, 2004) reflect only a small fraction of the actual present interest.

Reinforcement learning recently suggested the framework of *predictive state representations* (PSRs) (Littman, Sutton, & Singh, 2002; James & Singh, 2004). In PSRs, states are represented by the expectable predictions of the future in that state. Thus, states are not represented by themselves but rather by a collection of predictions about the future. Analyses showed that the representation is as powerful as POMDP models (hidden Markov models) and of size no larger than the corresponding POMDP model (Littman, Sutton, & Singh, 2002). Since POMDP-based model learning methods scale rather poorly, it is hoped that PSR-based learning methods may be learned faster and have better scalability properties (Singh, Littman, Jong, & Pardoe, 2003; James & Singh, 2004). However, at least the current implementations of PSRs appear hardly cognitive, requiring an explicit distinction of states, represented by a concatenation of predictions.

Observable operator models (OOMs) are an alternative approach to PSRs that currently appear to be slightly further developed, as indicated by their online learning capabilities and scalability (Jaeger, 2000; Jaeger, 2004). Observable operators are operator activities that reflect a certain event in the environment. Most recently, Jaeger (2004) proposed an online learning algorithm for learning the required set of suitable events and for learning their respective probabilities. Like PSRs, OOMs are an alternative approach to other POMDP models where OOMs are at least as expressive as POMDP models.

Could we be able to actually learn predictive representations in a much more distributed, LCS-like learning fashion? First approaches in that direction, also referred to as *anticipatory learning classifier systems* (ALCSs) were proposed elsewhere (Stolzmann, 1998; Butz, Goldberg, & Stolzmann, 2002).

The ACS2 system (Butz, 2002) showed able to learn in diverse problem domains. It can solve the multiplexer problem but is also able to handle predictions in diverse maze and blocks world scenarios. However, ACS2 depends
on a deterministic environment and is designed to learn complete predictions
of next states. Also, other similar learning systems such as YACS (Gérard &
Sigaud, 2001b; Gérard & Sigaud, 2001a) depend on determinism.

   The ALCS system MACS (Gérard & Sigaud, 2003) is modularized in that
different predictive modules predict different perceptual (or sensory) features
in the environment. This has the advantage of evolving separate modules for
potentially independent problem features. Currently, though, the modules are
learned completely independently so that knowledge exchange or knowledge
reuse is impossible. Since the behavior of problem features can be expected to
be correlated, an algorithm needs to be designed that detects, propagates, and
exchanges useful structure among the learning modules. In this way, distinct
(predictive) modules may reuse similar structures for distinct but correlated
predictions. Thus, the learning structure needs to be biased towards detecting
compound problem features that should be accessible to all related learning
modules.

   The learning mechanisms in XCS do not rely on determinism but showed
to work well also in stochastic problem settings. They are also well-suited to
ignore task-irrelevant attributes for which they require only linear additional
computational effort. Thus, an XCS-like predictive mechanism may be created
that predicts future features similar to other ALCSs.

   The envisioned system would be able to ignore irrelevant problem features
as well as handle noisy problem cases. XCS may be used as a modular, predictive learning architecture in which the prediction of the next state is evaluated
in terms of predictive accuracy. The exchange of knowledge may be realized
by enabling structural exchange over the boundaries of the different predictive modules. Figure 12.1 visualizes the proposed learning structure where
each XCS module learns to predict the behavior of one problem feature. The
predictive modules exchange dependency structures by the means of simple
crossover or the model-based dependency structures introduced in Chapter 7.

   In close vicinity to predictive learning lies the realm of associative learning.
The only difference to predictive learning is that in associative learning both
patterns are available at the same time and not in succession. Nonetheless,
associations may be learned by two interconnected predictive modules that
learn to predict the one pattern from the other and vice versa. This method
can be implemented easily with the proposed predictive extension of the XCS
classifier system. In comparison to neural network structures, the advantage
is the enhanced adaptivity gained by the evolutionary component in XCS
structuring connectivity and relevancy.

   More importantly for predictions and associations, the expected nature of
the interaction should be investigated before designing the actual associative
learner with its biases. It needs to be asked, which structures are expected
to be associated or which structural commonalities should be exploited. For

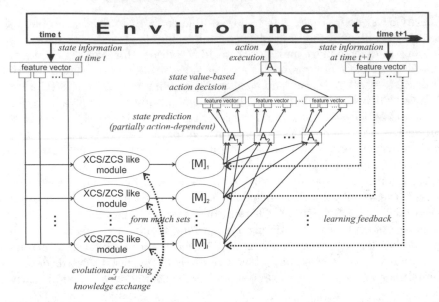

Fig. 12.1. The proposed modular, predictive learning architecture is well-suited for an XCS-based learning implementation. Predictions are formed in parallel. Learning is based on the difference between the predicted and the encountered environmental state information. Knowledge exchange between modules is accomplished by the means of genetic recombination of BB structures.

example, when predicting some sensory changes, higher activities of causes can be expected to induce stronger influences on the changes. On the other hand, when associating patterns, their dependency structures may strongly differ between modules so that different biases will be advantageous. Thus, an interactive pattern associator should be implemented that is predisposed to learn expected problem dependent structures. Statistical structure analysis, such as the ECGA or BOA mechanisms integrated in XCS, may be used to identify online which structures seem most suitable to predict the behavior of other structures.

12.2 Incremental Learning

Predictions cannot be known exactly beforehand so that learning of the predictions can only occur online, experiencing interactions between body and environment. Learning predictions may be predisposed or biased towards learning certain structures. Parts of these predispositions may already be realized by the provided sensory and motor capabilities. Thus, predictions can and should be biased towards learning certain structures and connections but need to be grown-up and adjusted to the actual body-related "hardware".

The idea of the importance of incremental learning is not new. Several previous promising systems can be identified, such as fast incremental learning in the ART networks (Grossberg, 1976a; Grossberg, 1976b) as well as initial LCS-like rule-based methods (Wilson, 1987b), to name a few early approaches.

The advantage of an incremental growth of a network structure, rather than fixed size and fixed connectivity from the beginning, is to direct computational needs to where they are really needed. Thus, growth is a critical time in development—as are the first few months in the life of an infant, shaping its mind for the rest of its life. However, growth is limited and once a distribution is reached that accomplishes current activity and tasks satisfactorily, it can be expected to become increasingly fixed. Growth takes time and energy and thus fights for limited resources. The resulting resource distribution, manifested in terms of neural concentration and connectivity, predisposes the cognitive structures. In the next learning stage, the existing connectivity may be shaped and finally only small adjustments may be possible.

What are the most desired features of a useful growth algorithm? Growth should be interactive, in that grown structure immediately influences the further structural growth. Growth should be modular, consisting of different, interdependent growth centers that have distinct representational responsibilities. The learning design should not only constrain the center's learning capability but also its connectivity and interactivity with other growth centers. These factors lead to a strong predisposition of the overall growing structure, which then learns in interdependent centers that detect and propagate structures, which are suitable for accurate predictions and consequent activity generation.

Growth needs to ensure covering, that is, the growing structure needs to assure that all potential inputs can be processed. Once the growth process is converging to a stable structure, no input or occurring state should result in a random reaction but should rather be influenced by all related experiences previously encountered by the growing cognitive system.

The essential question is how to create a growing, distributed structure mutually shaped and controlled by actual activity as well as by the (partially consequential) influences of the outside world. LCSs provide a solid framework that assures the evolution of a distributed knowledge representation. As we have seen in Chapter 10, the XCS system is comparable to a neural network learning structure but actually forms connectivity and detects relevancy online. While a neural network assumes relevancy of all inputs by default, XCS detects relevancy on the fly using evolutionary techniques to propagate and distribute successful relevancies. Thus, the XCS system might serve very well as the actual growing mechanism, evolving connectivity and thereby effectively shaping relevancy.

Again, the advantage of the accuracy-dependent learning enables XCS to be flexible and to shape its connectivity while further growing—reproducing successful structures mimicking duplication, mutating structures to search in the local-neighborhoods, recombining structures to detect relations and interdependencies. Once the structure is grown, other, more weight-oriented mechanisms might condense, shape, and refine the grown structure.

12.3 Hierarchical Architectures

Insights in cognitive and neural processing mechanisms suggest that many brain areas are structured hierarchically. Different but related areas usually interact bidirectionally. For example, the well-studied auditive system of humans and birds is hierarchically structured (Feng & Ratnam, 2000). Lower-level areas respond to specific sounds, such as phonemes or syllables, and higher levels use the extracted features responding to larger chunks of auditive input, such as words or song parts.

Even more prominently, vision research has shown that the visual cortex is structured modularly and hierarchically. Many distinct areas in the brain are responsible for feature-extraction and basic visual stimulus processing. These mechanisms work in parallel and are hardly influenced by the bottleneck of cognitive attention (Pashler, 1998). However, even these lower-level structure extraction mechanisms are strongly influenced by top-down processes such as attention related to object-properties, location properties, color properties, as well as predictive behavioral properties (Pashler, Johnston, & Ruthruff, 2001).

Few models exist that learn such hierarchical structures. Riesenhuber and Poggio (1999) introduce a hierarchical model for object recognition. By combining conjunctive and disjunctive feature extraction mechanisms, the representations in the hierarchy become progressively more focused on specific features and more generalized with respect to location or size. The behavior of the model is closely related to actual neuroscientific evidence. Giese and Poggio (2003) use the same principle to evolve a hierarchical movement recognition system that can distinguish various movement patterns. Again, the behavior of the model is closely related to actual experimental neuroscientific data of animals and humans. Poggio and Bizzi (2004) even suggest that such hierarchical pathways can be expected to exist not only in visual processing but also in motor control mechanisms. Interestingly, the models are all purely bottom-up, feed-forward systems. Although all articles suggest that top-down influences should be incorporated, none of them do.

An interesting approach that integrates bottom-up and top-down influences can be found in Rao and Ballard (1997). The model combines bottom-up, top-down and potentially feedforward signals by weighing them according to their signal-to-noise ratios, similar to the Kalman principle (Kalman, 1960; Haykin, 2001). The model evolves feature detectors typically found in the lower-layers of the visual cortex in animals (Rao & Ballard, 1999). Additionally, the model is capable of reconstructing previously seen images and identifying novel (parts of) images. However, so far the model has not been used to simulate actual top-down attention-related influences.

Several other more computer science-related hierarchical learning mechanisms exist. Hierarchical clustering forms hierarchical problem representations. However, regardless if the taken approach is agglomerative, in which many clusters are progressively combined, or divisive, in which one cluster is progressively divided, the clusters are usually non-overlapping in each level of the hierarchy (see e.g. Duda, Hart, & Stork, 2001 and citations therein). Hierarchical self-organizing maps (Dittenbach, Merkl, & Rauber, 2000) are available that grow hierarchies on demand. However, these hierarchies are also non-overlapping. Neither approaches consider reusing certain useful clusters in a lower level by several clusters on the higher level.

In the RL literature, hierarchical RL has gained increasing interest over the recent years (Sutton, Precup, & Singh, 1999; Dietterich, 2000; Drummond, 2002). Sutton, Precup, and Singh (1999) proposed the framework of *options*—actions extended in time—that can be used hierarchically. For example, in a problem with four interconnected rooms, an option might learn the way to the doorway. This option may then be combined with other options effectively learning to cross several rooms to reach a reward position much more effectively. Similarly, the hierarchy may be used to evolve a predictive representation of the environment. Barto and Mahadevan (2003) give an excellent overview of the current state-of-the-art in hierarchical reinforcement learning. While all papers emphasize the importance of automatically learning such hierarchies, most of the papers focus on how to apply options but not how to learn a hierarchy in the first place. Most recently, it was suggested to learn the hierarchy on the fly by detecting *transitional states* in the explored environment applying simple statistical methods monitoring the temporal state distribution online (Butz, Swarup, & Goldberg, 2004; Simsek & Barto, 2004).

Clearly, neural structures including the multi-layer perceptron are reusing the representations emerging in hidden layer units for classification or prediction on the higher level. Even more potential seems to lie in the recently emerging multi-layer kernel-based representations in which kernel-based feature extraction units evolve in the lower-layer, which are then combined on the higher layer for the task at hand.

One of the simplest of these networks is the radial-basis function network (RBF), which evolves radial-basis kernels approximating the proximity of the actual data to each of the hidden units (see e.g. Hassoun, 1995 and citations therein). Other kernels are expected to be useful for other types of problems with other types of non-linearities, that is, other types of dependency-structures, or BBs, in the problem space. An excellent overview of which kernels may be suitable for which types of problems and expectable patterns can be found in Shawe-Taylor and Cristianini (2004).

The XCS framework is ready to serve as the module that detects relevant kernel structures, growing them problem dependently. The real-valued problem analysis in Chapter 9 has shown that XCS can detect and evolve patterns in many problem domains. Thus, XCS is readily combinable with kernel-based feature extraction mechanisms in which the feature vector could consist of a set of kernels that evolve with the classifier population over time.

Another, more hierarchical approach is imaginable in which lower-level structures evolve interactively with higher level structures. In fact, lower-level kernels may be suitable to be reused in many higher-level classification rules, function approximation neurons, or predictor units. Hierarchical structures are imaginable in which the upper levels reward the lower levels for providing useful structure activity.

Since XCS is designed to learns structural patterns online solely dependent on reward-based feedback, it appears to be a very suitable learning mechanism for each of the levels. Figure 12.2 shows what such a hierarchical XCS structure could look like. Activity is propagated bottom-up. Each layer is predisposed by the top-down activity to process bottom-up activity (attention). Thus, subsequent layer activity is predisposed by the inner state of the system but ultimately determined by the bottom-up input. Note that the architecture certainly does not need to be strictly bottom-up, top-down biased. Essentially, it is imaginable that the bottom-up influence comes from other activity centers. It might not necessarily stem from sensory activity.

Upper layers provide feedback in terms of reinforcement, rewarding lower layers for their activity. Lower layers compete for this feedback, consequently evolving a complete problem distribution clustered in a way suitable for higher level predictions, classifications or whatever the task might be. Additional layer-owned feedback may be provided to ensure a base-rate activity in each layer.

The advantage of the solely reinforcement-based learning is that feedback might come from any other layer so that the shaping of one layer may be coordinated by the needs of whatever layer is interested in the evolving information. This enables the designer to strongly bias the learning structure by defining neighborhoods of layers that influence each other to a degree controlled by the neighborhood structure.

We recently experimented with a related structure that searches for hierarchical language patterns over time (Butz, 2004b). The proposed cognitive sequence learning architecture (COSEL) could detect most frequent characters, syllables, and words in a large text document, growing a hierarchical problem representation from scratch. While COSEL is still in its infancy, growing hierarchical architectures in which the top-down influence shapes the evolution of the bottom-up growth, stabilizing and guiding it, might be one of the key elements to learn stable, distributed, hierarchical cognitive learning architectures.

12.4 Anticipatory Behavior

Over the last decades, psychology realized that most goal-directed behavior is actually not purely stimulus-response driven but strongly influenced by the

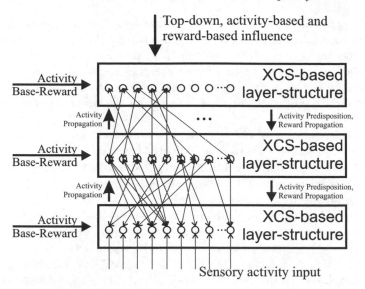

Fig. 12.2. The proposed hierarchical XCS-based learning architecture propagates neural activity bottom-up and predisposes the propagation top-down. Learning is purely reward-based and depends on the appropriateness of the activity predispositions and the interdependent activity structure.

expectations about the future (Tolman, 1932; Hoffmann, 1993; Hoffmann, 2003; Hommel, Müsseler, Aschersleben, & Prinz, 2001). It was shown that anticipations influence behavioral selection, initialization, and execution (Kunde, Koch, & Hoffmann, 2004). Moreover, anticipation guides attentional processes (Pashler, Johnston, & Ruthruff, 2001). Finally, anticipation is responsible for higher level cognition, potentially leading to self-controlled consciousness in terms of a predictive, attention-based control system (Taylor, 2002).

For a more detailed review on anticipatory behavior including its manifestations in different research disciplines as well as its potential for the development of competent cognitive architectures, the interested reader is referred elsewhere (Butz, Sigaud, & Gérard, 2003; Butz, 2004a). Clearly, however, predictions, associations, incrementally growing structures, and hierarchies are some of the fundamental building blocks of anticipatory behavior and cognition.

Prediction is that part of an anticipation that predicts future changes, states, and progress. Anticipatory behavior is then the part that decides on how the predictions actually influence the cognitive system. As mentioned before, associations are similar to predictions and similarly serve as structures that enable anticipatory behavior.

Hierarchical structures enable anticipations on multiple levels of abstraction in time and space. Starting from very basic sensory inputs on the lowest level, a hierarchical representation needs to develop in which higher levels represent progressively larger and more abstract representations of the environment. Each level may serve as a predictor for the lower level and as an information propagator to the higher level. In that way, each level predicts the next lower level input given its current activity, resulting in an anticipatory behavioral influence on the lower level. In the meantime, the activity will be propagated to the next higher level, influencing the resulting next prediction from that level.

The hierarchical abstraction can be accomplished in space and time. In space, the abstraction leads to progressively larger and more abstract objects or entities in the environment. In time, on the other hand, the hierarchy leads to longer sequences in time. In relation to speech, for example, the hierarchy could abstract from phonemes to syllables to words to whole sentences. In relation to face recognition, the hierarchy could abstract from edges, corners, colors, shades, and other basic visual features to eyes, nose, mouth, ears, and so forth to relations between them to overall relations between those relations (see Riesenhuber & Poggio, 2000; Giese & Poggio, 2003 for related efforts). Each higher-level abstract representation allows the prediction of lower level relations—as the imagination of a face facilitates the detection of that face or the expectation of a train of thought facilitates word recognition.

Thus, the hierarchies enable anticipatory processes in time and space having high potential for explaining attentional phenomena but also the mentioned influences on action decision, initiation, and execution. Moreover, learning is influenced by this top-down influence since only activity can cause learning and the top-down controlled predisposition of activity consequently predisposes learning.

It remains to be shown how these hierarchical levels may be interconnected and how they might be grown guided by its own activity and the environmental influences. The XCS framework provides a potentially valuable tool to model levels in the hierarchy. Since the system is only dependent on reinforcement feedback, higher levels may reward lower-level activity that propagates activity successfully. Classifiers serve as activity propagation entities to the next higher level. Even more importantly, they also serve as the predictors of lower level activity. Accurate predictors will gain more reward and thus establish themselves in the hierarchy. Anticipations manifest themselves in higher level classifier activity that predisposes lower level activity.

12.5 Summary and Conclusions

We discussed four fundamental building blocks of cognitive learning structures including predictive and associative representations, incremental learning, hierarchical learning, and anticipatory behavior. The first three types basically enable the fourth type, which in turn influences and stabilizes the development of the former three. Thus, the four discussed aspects of cognition are highly interactive, passing reward and activity information that cause connective attraction and repulsion, growth and death.

The XCS/LCS framework can be extended to face the design challenges of such a highly interactive, cognitive learning system. However, the successful design of such an architecture is not straightforward. Simple, one dimensional top-down feedback may not be enough to develop, propagate, and grow substructures successfully. The modular and hierarchical structure and connectivity between modules remains to be developed in detail. The XCS framework, nonetheless, appears to have the potential to serve as the tool that shapes the proposed hierarchically structures modules interactively—adapting to continuous input and feedback while filtering noisy and partially misleading learning signals.

13

Summary and Conclusions

This book has investigated rule-based evolutionary online learning systems, often referred to as learning classifier systems (LCSs). The proposed facetwise problem analysis approach showed that the XCS classifier system is an effective online learning and generalizing predictive reinforcement-based system that evolves a complete, maximally accurate and maximally general problem solution quickly and reliably. The scalability analysis showed that XCS can PAC-learn restricted k-DNF functions. However, XCS is not restricted to the world of k-DNF problems but can also learn in real-valued problem domains, regardless if the task is to classify data or to approximate functions. Moreover, XCS can be applied in multistep reinforcement learning (RL) problems, in which the system was shown to learn optimal behavioral policies effectively filtering noise and ignoring irrelevant attributes. The extension of the proposed facetwise analysis approach for LCS-based system design suggested strong potentials for integrating LCS systems and LCS-based learning components into modular, hierarchical, predictive learning structures.

In the following, we summarize the major findings and achievements of this book. Next, we provide conclusions focusing on learned lessons for system analysis and design and on future successful LCS-driven applications.

13.1 Summary

This book showed that LCSs can be successfully applied in classification problems, function approximation problems, and RL problems. From a more general perspective, LCSs can be applied in any problem in which accurate predictions need to be learned. LCSs evolve distributed problem solutions in which different subsolutions apply in different problem subspaces. Since RL problems only provide reinforcement feedback, which can be potentially delayed, reward needs to be estimated and propagated most effectively to prevent learning disruption due to inaccurate reward distributions.

All relevant problem domains require an effective search through potential problem solution structures. Usually, solution structure can be expected to contain lower-level structures, referred to as building-blocks (BBs), that guide to the maximally accurate problem solution. Gradient-based techniques are used for appropriate structure evaluation. RL-based techniques, such as Q-learning, ensure effective reward distribution. Genetic algorithms (GAs) propagate and recombine BBs and consequently inductively evolve, or grow, the desired distributed problem solution. The solution is represented by a population of rules, or classifiers. Subsets of rules are responsible for different problem subspaces.

Facetwise LCS Approach

Our facetwise LCS theory approach suggests that the following four major aspects need to be satisfied to assure a successful application of LCSs:

1. Evolutionary pressures need to apply appropriately.
2. Solution growth and sustenance need to be enabled and ensured.
3. Solution search needs to be effective.
4. Multistep problems pose additional challenges concerning reward propagation and environmental exploration.

In particular, evolutionary pressures need to ensure that classifier parameter estimations *guide* towards better solutions and ultimately lead to the optimal solution disabling overgenerals. *Generalization pressure* needs to apply to favor most compact solutions and *fitness pressure* needs to propagate solutions that yield higher fitness. Hereby, fitness pressure may not be overruled by generalization pressure. *Parameter estimates* need to be approximated as fast and accurate as possible but also need to be continuously adjusted to population dynamics, where necessary. Fitness overestimations in young and inexperienced classifiers need to be avoided.

To ensure solution growth, *evaluation time* needs to be available to initiate the evolutionary process. Secondly, the *supply* of minimal order classifier representatives needs to enable the identification of better classifiers. Additionally, time for *identification and reproduction* needs to assure the accurate identification of better classifiers as well as the growth of these better classifiers. Once growth is assured, *solution sustenance* requires the implementation of effective niching techniques such as fitness sharing, niche reproduction, and restricted replacement.

To ensure the fastest search for the problem solution, mutation needs to result in an *effective neighborhood search* in the vicinity of current best subsolutions. Recombination needs to combine important subsolution structures to guarantee quick learning progress. Hereby, *global* structural commonalities and *local* structural differences need to be detected and properly considered in the recombination mechanism.

In *multistep* RL problems, distortions in *niche occurrence frequencies*, proper *behavioral policies*, and *accurate reward propagation* need to be additionally considered. Due to the interdependence of RL-based rule evaluation and GA-based rule evolution, reward needs to be propagated most accurately. The implemented behavioral policy may partially prevent niche occurrence frequency distortions.

The XCS Classifier System

With the facetwise approach in hand, the XCS classifier system was introduced and analyzed in detail. XCS is an accuracy-based LCS that is designed to reliably evolve a complete, maximally accurate, and maximally general problem solution. The RL component in XCS utilizes adapted Q-learning mechanisms and other gradient-based principles to accurately estimate rule reward prediction, prediction error, and fitness. The GA component is a steady-state niche GA that reproduces, mutates and recombines classifiers in problem niches (action sets) and deletes classifiers from the whole population.

The subsequent evolutionary pressure analysis showed that XCS applies an *implicit semantic generalization* pressure, combining niche-based reproduction with population-wide deletion. The derived *specificity equation* quantifies this pressure, taking the specificity effect of simple mutation into account. *Fitness pressure* is the main drive towards maximal accuracy. *Tournament selection* with tournament sizes proportional to the current niche size assures reliable fitness pressure with a constant, minimal strength that depends on the tournament size proportion. *Subsumption deletion* provides further *syntactic generalization* pressure in deterministic problems.

The analysis of solution growth and sustenance revealed that XCS is a machine learning competitive learning system whose learning complexity scales polynomially in solution complexity, problem length, learning reliability, and solution accuracy. The order of the polynomial depends on the minimal order of problem difficulty k_m, which denotes the minimal number of specified attributes necessary to decrease class distribution entropy.

As in GAs, in LCSs lower order BB structure may need to be identified, propagated, and recombined effectively. In XCS, BBs are subsets of specified attributes that increase accuracy only as a whole. This is the case in hierarchically structured classification problems in which substructures are evaluated independently and then combined in an overall solution evaluation mechanism. To make the evolutionary search more effective in such BB-hard classification problems, statistical structure analysis methods such as Bayesian networks can replace recombination. However, in contrast to the incorporation of such mechanisms in GAs, in LCSs local structural subsolution differences need to be accounted for. Thus, the statistical dependency models need to be adjusted to reflect local problem properties in order to yield niche-based statistical offspring generation mechanisms.

The subsequent analysis in binary classification problems confirmed the theoretic analysis. We showed that XCS performance is robust to noisy reward feedback and noisy problem instances. The investigated problems included large multiplexer problems, combined multiplexer problems that partially require appropriate structure identification and propagation, problems that require an overlapping problem solution representation, and problems in which different problem subspaces require different subsolution complexities.

Also, applications in multi-valued datamining problems confirmed XCS's competence. Comparing XCS's performance in forty-two datasets with ten other machine learning techniques yielded competitive performance scores. In fact, no machine learning method was found that consistently outperformed XCS and XCS did not consistently outperform any other learning method (except majority). The results confirm the expectable performance differences due to the different learning biases in each machine learner and the different structural properties of each dataset. The application to function approximation problems further confirmed the strong predictive capabilities of XCS. The applications showed that other condition structures (such as radial-basis conditions) as well as other predictive approximations (such as linear predictions) can be incorporated easily into the general XCS learning architecture.

In all these evaluations, it needs to be kept in mind that XCS learns a problem solution online receiving only scalar (reinforcement) feedback. Thus, the learning mechanism in XCS is very general, searching online for relevant problem substructures and maximally accurate predictions in the identified substructures. This hypothesis was confirmed in Chapter 10, in which XCS was evaluated on diverse RL problems, including maze and blocks world scenarios. Realizing the connection to neural-based RL techniques, we introduced gradient-based updates to stabilize reward estimation and propagation. Hereby, we noticed that XCS combines the neural-based reward estimation approach with a somewhat tabular-based structure identification mechanism. The GA component searches for solution subspaces that yield similar reward values, whereas the reward component evaluates the evolved subspaces. This is why residual gradient approaches appear unnecessary in the current XCS system. As a whole, the performance analysis showed that XCS is able to solve a variety of multistep RL problems effectively. Irrelevant, randomly changing problem features require only low-order (often linear) polynomial computational effort, dependent on the solution complexity and the size of the problem space.

Facetwise LCS Design

The successful facetwise XCS analysis suggests that the facetwise approach can be carried over to analyze other LCSs and, even more importantly, to design new, advanced LCS architectures based on the gained insights. Chapter 11 reviewed the major LCS learning mechanisms and contrasted their influences on learning performance with respect to the facetwise theory aspects.

Besides the learning theory aspects, we also saw that the chosen condition structure as well as the chosen prediction mechanism strongly influence performance. The power of an LCS-based learning approach is that its structural constrains are very flexible. Different condition structures as well as different predictive representations can be incorporated and combined easily. The facetwise theory can be adjusted to the chosen representation at hand so that the general LCS theory holds for any representations chosen for conditions and predictions.

Consequently, LCSs provide a toolbox of learning mechanisms and structural representations that can be suitably combined for the problem at hand. The insights should be used to design new, advanced LCS-based learning structures, which are designed to solve particularly important problem structures most effectively.

For the design of cognitive learning structures, modular, hierarchical, predictive learning architectures seem most effective. Due to the flexibility of LCS systems and their predictive capabilities, each learning module in such a cognitive system may be realized by an LCS-based mechanism. Moreover, since cognitive learning systems require online learning mechanisms—which potentially only receive sporadic reward feedback—LCSs have the potential to serve as an important entity for the successful design of such effective, interactive, adaptive systems.

13.2 Conclusions

As shown, a facetwise analysis approach enables the successful analysis of highly interactive learning systems such as the XCS classifier system. We were not only able to qualitatively understand the system, but we were also able to quantify learning behavior and computational learning requirements. Moreover, the facetwise approach enabled modular system design and system improvement, as shown with the introduction of tournament selection, statistics-based search techniques, gradient-based update techniques, and advanced condition as well as predictive representations. Thus, the facetwise approach enables flexibility and robustness in system analysis as well as in system design.

Which are the most promising future directions for LCS-based learning systems? Clearly, the analysis of other LCS-based systems is a first priority. Besides the analysis of existing systems, the enhancement of existing systems, guided by the facetwise approach, as well as the design of novel systems promises to yield immediate payoff. Hereby, it seems most important to target enhancements and modifications towards specific problem structures. If structures are properly understood, structure-specific enhancements can be expected to be maximally effective.

To realize such enhancements, it is consequently necessary to understand first the underlying problem structure, the available feedback, and the intended solution structure. The understanding of problem structure and expected problem solution structure leads to the design of an appropriate problem solver. Learning needs to be biased towards searching for the expected structures.

For example, if the solution structure can be expected to be conditionally linearly dependent on problem input, conditionally linear predictions can be expected to be most effective. The more suitable the condition structure to differentiate the conditional subspaces, the higher accurate solutions are expected to evolve.

In the case of a piecewise solution approximation, two issues need to be considered: (1) How may the space be partitioned most effectively? (2) What kind of predictive structure is expected to be relevant in each of the partitions? For example, if the partitioning is expected to be linear in the problem features, linear decision-theoretic partitions will be most appropriate, as done in the incremental model tree learner introduced in Potts (2004). Similarly, if subsolutions can be expected to be linear transformations, linear solution transformations and learners can be expected to be most effective.

With the knowledge of addressed problem structure, solution structure, solution representation, and biased learner at hand, it is then possible to determine the competence of the learner with respect to its computational requirements and scalability. In essence, it is necessary to identify how much computational effort is necessary to evolve subspace representations and corresponding predictive solution structures. The facetwise theory approach for LCSs specifies how to conduct these derivations in LCS-based learning systems.

Expected structural similarities, and thus subsolution structure exchange, can be expected to be helpful in the evolution of a complete problem solution. For example, a multi-layer neural network strongly assumes structural similarities in different problem subspaces, reusing hidden layer representations for the derivation of problem solutions. A tabular approach, such as tabular Q-learning, assumes no structural similarities, developing independent solution representations for each problem instance (such as each possible state-action pair in an MDP). An XCS-based approach exchanges structural information to identify similarly structured problem subspaces relevant for different problem solutions. The enhanced offspring generation mechanism in XCS, which uses statistical tools to analyze currently most effective solution structures, biases local offspring generation towards global structural dependencies. It is consequently helpful in problems in which similar solution structures are present in different problem subspaces (as exemplified by the investigated hierarchical, BB-hard problems).

Besides these important system considerations and potential system enhancements, it is necessary to consider in which problems LCS-based learning mechanisms may be applied most effectively. This leads to the reconsideration of the strengths of LCS-based systems.

As we have seen, LCSs are effective *online learning* systems that interactively learn a *distributed problem solution*. If online learning is not required, or the solution is expected to be non-distributed, LCSs may not be necessary to solve the problem. The distributed problem solution in LCSs partitions the problem space into potentially overlapping problem subspaces in which particular problem solutions apply. Thus, problems in which conditionally independent subsolutions need to be learned are most suited for LCSs. Moreover, the nature of the subsolution representation is an important consideration. We showed that at least constant and linear subsolution representations are possible. Lanzi, Loiacono, Wilson, and Goldberg (2005) showed that polynomial representations are also possible. Finally, if the subspaces have similar properties, the evolutionary learning algorithm can be expected to be most effective in searching for the effective subspaces. If lower order BBs are expected to exist in multiple subspaces, then effective BB identification and processing mechanisms can be expected to improve evolutionary search even further.

Although the capabilities of LCS-based learning systems appear quite general, one application of LCS systems seems to be extraordinarily appealing. As envisioned by John Holland in his original publication of the cognitive system CS1 (Holland, 1976; Holland, 1977), LCSs seem very suitable to be successfully applied in cognitive learning structures. Hereby, the LCS mechanisms may serve as the main processes that grow highly interactive, cognitive learning structures solely guided by reward and activity information exchanged between interactive LCS-based learning modules. The structural flexibility of the systems enables the generation of potentially necessary enhanced rule structures that result in the generation of advanced interactive, predictive patterns. With the facetwise approach to LCS analysis and design, as well as with the representational understanding at hand, LCS mechanisms may soon serve as the building blocks for the successful design of highly modular and hierarchical predictive cognitive learning systems.

A

Notation and Parameters

Problem Notation

Concept Learning (Classification) Problems

$\mathcal{S}$ problem space
$s \in \mathcal{S}$ actual problem instance
$\mathcal{S} = \{0, 1\}^l$ binary problem space of length l
l problem length l (that is, number of features in a problem instance)
$\mathcal{A}$ problem classes
$a \in \mathcal{A}$ actual problem class
$\mathcal{A} = \{0, 1\}$ binary (two-class) problem
n number of problem classes

Reinforcement Learning (MDP) Problem

$\mathcal{S}$ set of possible sensory inputs
$s \in \mathcal{S}$ current sensory input
l number of sensory features
$\mathcal{A}$ available actions
$a \in \mathcal{A}$ actual action
n number of possible actions
$\mathcal{R}$ reward feedback
$r \in \mathcal{R}$ current reward

Problem Difficulty

k_m minimal number of attributes necessary to decrease entropy of class distribution
k_d order of problem difficulty—the target concept space is exponential in k_d

Simple Learning Classifier System

System Parameters
N maximal population size
ϵ ϵ-greedy strategy parameter
γ reward discount factor
β learning rate
$P_\#$ don't care probability
μ probability of mutating a condition attribute (or the action)
χ probability of applying the chosen crossover operator

Classifier Parameters
C condition part; in binary problems, $C \in \{0, 1, \#\}^l$
A action part $A \in \mathcal{A}$
R reward prediction

Other Notations
$[P]$ classifier population
$[M]$ match set
$[A]$ action set
$\sigma(cl)$ specificity of condition part of classifier cl; in binary problems,
       specificity is defined as the number of specified attributes divided by l
$\sigma[X]$ average specificity in classifier set X

XCS Classifier System

LCS Parameters

N maximal population size

$P_\#$ don't care probability

P_I, ε_I, F_I default parameter initialization values set to 500, 500, and .01, resp.

ϵ ϵ-greedy strategy parameter

γ reward discount factor

β learning rate

$\alpha, \varepsilon_0, \nu$ accuracy determination parameters

θ_{GA} threshold that controls GA invocation

τ the tournament size (proportion of current action set size)

μ probability of mutating a condition attribute (or the action)

χ probability of applying the chosen crossover operator

θ_{del} threshold that requires min. experience for fitness influence in deletion

δ fraction of mean fitness below which deletion probability is increased

θ_{sub} threshold that requires minimal experience for subsumption

Classifier Parameters

C condition part; in bin. problems, $C \in \{0, 1, \#\}^l$; in real-val. problems, hyperrectangles are represented by $C = (l_1, u_1, l_2, u_2, ..., l_l, u_l)$, hyperspheres by $C = (m_1, m_2, ..., m_l, \sigma_*)$, hyperellipsoids by $C = (m_1, m_2, ..., m_l, \sigma_1, \sigma_2, ..., \sigma_l)$, and general hyperellipsoids by $C = (m_1, m_2, ..., m_l, \sigma_{1,1}, \sigma_{1,2}..., \sigma_{n,n-1}\sigma_{n,n})$.,

A action part $A \in \mathcal{A}$

R reward prediction

ε mean absolute reward prediction error

F fitness (in macro classifiers)

as the mean action set size the classifier is part of

ts last time stamp the classifier was part of a GA application set

exp the number of evaluation steps the classifier underwent so far

num the numerosity, that is, the number of micro-classifiers represented by this (macro-) classifier

Other Notations

$[P]$	classifier population
$[M]$	match set
$[A]$	action set
$\sigma(cl)$	specificity of condition part of classifier cl; in binary problems, specificity is defined as the number of specified attributes divided by l
$\sigma[X]$	average specificity in classifier set X
κ	current accuracy of a classifier
κ'	current set-relative accuracy of a classifier
ρ	(combined) reward received by the current action set
$P(\mathcal{A})$	prediction array estimating the value of each possible action
$P(A)$	estimated predictive value of action A
$\Delta_{mf}(\sigma)$	Average change in specificity caused by free mutation
$\Delta_{mn}(\sigma)$	Average change in specificity caused by niche mutation
δ_P	learning error probability, that is, probability of learning failure
ε_P	solution inaccuracy, i.e., incorrectly classified problem subspace prop.
θ_{be}	minimum experience exp required for model building classifiers
θ_{bn}	minimum numerosity num required for model building classifiers
$\theta_{b\varepsilon}$	maximum error ε allowed in model building classifiers

B

Algorithmic Description

This section gives a concise algorithmic description of the XCS classifier system. A similar description can be found elsewhere (Butz & Wilson, 2001; Butz & Wilson, 2002). It differs in the offspring classifier initialization technique as well as the subsumption technique.

The algorithmic description uses the dot notation to refer to classifier parameters. Sub-functions are indicated by pure capital letters. We use indentation to indicate the length of a sub-clause such as the effect of an if statement or a for loop.

Overall Learning Iteration Cycle

At the beginning of a run, XCS's parameters may be initialized, iteration time needs to be reset, and problem may be loaded / initialized. XCS's population $[P]$ may be either left empty or may be initialized with the maximal number of classifiers N, generating each classifier with a random condition and action and initial parameters. The two methods usually differ only slightly in their effect on performance. Moreover, classifier generation due to covering assures that the problem space is immediately covered with respect to the problem space distribution, resulting in an additional advantage. After XCS and problem initializations, the main learning loop is called.

```
XCS:
 1 initialize problem env
 2 initialize XCS
 3 RUN EXPERIMENT
```

In the main loop *RUN EXPERIMENT*, the current situation or problem instance is first sensed (received as input). Second, the match set is formed from all classifiers that match the situation. Third, the prediction array $P(A)$ is formed based on the classifiers in the match set. Based on $P(A)$, one action is chosen and the action set $[A]$ is formed. Next, the winning action is executed. Then the previous action set $[A]_{-1}$ (if this is a multistep problem and there is a previous action set) is modified using the previous reward ρ_{-1} and the largest action prediction in the prediction array. Moreover, the GA may be applied to $[A]_{-1}$. If a problem ends on the current time-step (single-step problem or last step of a multistep problem), $[A]$ is modified according to the current reward ρ and the GA may be applied to $[A]$. The main loop is executed as long as the termination criterion is not met. A termination criterion is, e.g., a certain number of trials or a persistent 100% performance level. Here we stick to a fixed number of trials.

RUN EXPERIMENT():

```
 1 while(termination criteria are not met)
 2     s ← env: get situation
 3     GENERATE MATCH SET [M] out of [P] using s
 4     GENERATE PREDICTION ARRAY P(A) out of [M]
 5     a ← SELECT ACTION according to P(A)
 6     GENERATE ACTION SET [A] out of [M] according to a
 7     env: execute action a
 8     r ← env: get reward
 9     if([A]_-1 is not empty)
10         ρ ← r_-1 + γ * max P(A)
11         UPDATE SET [A]_-1 using P possibly deleting in [P]
12         RUN GA in [A]_-1 considering s_-1,ρ inserting in [P]
13     if(env: eop)
14         ρ ← r
15         UPDATE SET [A] using P possibly deleting in [P]
16         RUN GA in [A] considering s,ρ inserting in [P]
17         empty [A]_-1
18     else
19         [A]_-1 ← [A]
20         r_-1 ← r
21         s_-1 ← s
```

Sub-procedures

The main loop specifies many sub-procedures essential for learning in XCS. Some of the procedures are more or less trivial while others are complex and themselves call other sub-procedures. This section describes all procedures specified in the main loop. It covers all relevant processes and describes them algorithmically.

Formation of the Match Set

The *GENERATE MATCH SET* procedure receives the current population $[P]$ and the current problem instance s as input. While matching is rather trivial, covering may take place. Covering is called when the number of different actions represented by matching classifiers is less than the parameter θ_{mna} (this is usually set to the number of possible classifications n in a problem). Thus, *GENERATE MATCH SET* first looks for the classifiers in $[P]$ that match s and next, checks if covering is required. Note that a classifier generated by covering can be directly added to the population since it differs from all current classifiers.

GENERATE MATCH SET([P], σ):
```
 1 initialize empty set [M]
 2 for each classifier cl in [P]
 3     if(DOES MATCH classifier cl in situation s)
 4         add classifier cl to set [M]
 5 while(the number of different actions in [M] < θ_mna)
 6     GENERATE COVERING CLASSIFIER cl_c considering [M]
                                        and covering s
 7     add classifier cl_c to set [P]
 8     DELETE FROM POPULATION [P]
 9     add classifier cl_c to set [M]
10 return [M]
```

The following paragraphs describe the sub-procedures included in the *GENERATE MATCH SET* algorithm.

Classifier Matching

The matching procedure is commonly used in LCSs. A 'don't care'-symbol # in C matches any symbol in the corresponding position of s. A 'care', or non-# symbol, only matches with the exact same symbol at that position. This description focuses on binary problems and thus a ternary alphabet for XCS's conditions. Note that classifier conditions may be expressed more efficiently by specifying the care-positions only. In this way, the matching process can be sped-up significantly since only the specified attributes are relevant. If all comparisons hold, then the classifier matches s and the procedure returns *true*.

DOES MATCH(cl, s):
```
 1 for each attribute x in cl.C
 2     if(x ≠ # and x ≠ the corresponding attribute in s)
 3         return false
 4 return true
```

Covering

Covering occurs if there are less than θ_{mna} actions are represented by classifiers in $[M]$. If covering is triggered, a classifier is created whose condition matches s generating don't care symbols in the condition part with probability $P_\#$. The classifier action is chosen randomly from among those not present in $[M]$.

> *GENERATE COVERING CLASSIFIER([M], σ):*
> ```
> 1 initialize classifier cl
> 2 initialize condition cl.C with the length of s
> 3 for each attribute x in cl.C
> 4    if(RandomNumber[0,1[ < P#)
> 5       x ← #
> 6    else
> 7       x ← the corresponding attribute in s
> 8 cl.A ← random action not present in [M]
> 9 cl.P ← pI
> 10 cl.ε ← εI
> 11 cl.F ← fI
> 12 cl.exp ← 1
> 13 cl.ts ← actual time t
> 14 cl.as ← 1
> 15 cl.num ← 1
> ```

The Prediction Array

After the generation of the match set, the prediction array $P(A)$ provides reward estimates of all possible actions. The *GENERATE PREDICTION ARRAY* procedure considers each classifier in $[M]$ and adds its prediction multiplied by its fitness to the prediction value total for that action. The total for each action is then divided by the sum of the fitness values for that action to yield the system prediction.

> *GENERATE PREDICTION ARRAY([M]):*
> ```
> 1 initialize prediction array P(A) to all null
> 2 initialize fitness sum array FSA to all 0.0
> 3 for each classifier cl in [M]
> 4    P(cl.A) ← P(cl.A) + cl.P * cl.F
> 5    FSA(cl.A) ← FSA(cl.A) + cl.F
> 6 for each possible action A
> 7    if(FSA(A) is not zero)
> 8       P(A) ← P(A) / FSA(A)
> 9 return P(A)
> ```

Choosing an Action

XCS is not dependent on one particular action-selection method, and any of a great variety can be employed. For example, actions may be selected randomly, independent of the system predictions, or the selection may be based on those predictions—using, e.g., proportionate selection or simply picking the action with the highest system prediction. However, note that a non-random action selection algorithm biases niche occurrence and may thus require additional population size adjustments.

In our *SELECT ACTION* procedure we illustrate a combination of *pure exploration*—choosing the action randomly—and *pure exploitation*—choosing the best one. This action selection method is known as *ε-greedy* selection in the reinforcement learning literature (Sutton & Barto, 1998). Due to the mentioned distribution effect, XCS learns more stable if the GA is applied in exploration steps only.

```
SELECT ACTION(PA):
1 if(RandomNumber[0,1[ > ε)
2     //Do pure exploitation here
3     return the best action in P(A)
4 else
5     //Do pure exploration here
6     return a randomly chosen action
```

Formation of the Action Set

After the match set $[M]$ is formed and an action is chosen for execution, the *GENERATE ACTION SET* procedure forms the action set out of the match set. It includes all classifiers that propose the chosen action for execution.

```
GENERATE ACTION SET([M], a):
1 initialize empty set [A]
2 for each classifier cl in [M]
3     if(cl.A = a)
4         add classifier cl to set [A]
```

Updating Classifier Parameters

The reinforcement portion of the update procedure follows the pattern of Q-learning (Sutton & Barto, 1998). Classifier predictions are updated using the immediate reward and the discounted maximum payoff anticipated on the next time-step. Note that in single-step problems, the prediction is updated using only the direct reward ρ.

Each time a classifier enters the set $[A]$, its parameters are modified in the order: exp, ε, p, f, and as. Hereby, the update of the action set size estimate is independent from the other updates and consequently can be executed at any point in time. While the updates of exp, p, ε, and as are straightforward, the update of f is more complex and requires more computational steps. Thus, we refer to another sub-procedure. Finally, if the program is using action set subsumption, the procedure calls the *DO ACTION SET SUBSUMPTION* procedure.

```
UPDATE SET([A], P, [P]):
1  for each classifier cl in [A]
2      cl.exp++
3      //update prediction error cl.ε
4      if(cl.exp < 1/β)
5          cl.ε ← cl.ε + (|P - cl.P| - cl.ε) / cl.exp
6      else
7          cl.ε ← cl.ε + β * (|P - cl.P| - cl.ε)
8      //update prediction cl.P
9      if(cl.exp < 1/β)
10         cl.P ← cl.P + (P - cl.P) / cl.exp
11     else
12         cl.P ← cl.P + β * (P - cl.P)
13     //update action set size estimate cl.as
14     if(cl.exp < 1/β)
15         cl.as ← cl.as + (∑_{c∈[A]} c.num - cl.as) / cl.exp
16     else
17         cl.as ← cl.as + β * (∑_{c∈[A]} c.num - cl.as)
18 UPDATE FITNESS in set [A]
19 if(doActionSetSubsumption)
20     DO ACTION SET SUBSUMPTION in [A] updating [P]
```

Fitness Update

The fitness of a classifier in XCS is based on the relative *accuracy* of its reward predictions. The *UPDATE FITNESS* procedure first calculates the classifier's accuracy κ using the classifier's prediction error ε. Then the classifier's fitness is updated using the *normalized accuracy* computed in lines 8-10.

UPDATE FITNESS([A]):

```
 1  accuracySum ← 0
 2  initialize accuracy vector κ
 3  for each classifier cl in [A]
 4      if(cl.ε < ε₀)
 5          κ(cl) ← 1
 6      else
 7          κ(cl) ← α * (cl.ε / ε₀)⁻ᵛ
 8      accuracySum ← accuracySum + κ(cl) * cl.num
 9  for each classifier cl in [A]
10      cl.F ← cl.F + β( κ(cl) * cl.num / accuracySum - cl.F)
```

The Genetic Algorithm in XCS

The final sub-procedure in the main loop, *RUN GA*, is also the most complex. First, the action set is checked to see if the GA should be applied at all with respect to the θ_{GA} threshold. Next, two classifiers (i.e. the parents) are selected and reproduced. After that, the resulting offspring are possibly crossed and mutated. If the offspring are crossed, their prediction values are set to the average of the parents' values. Finally, the offspring are inserted in the population, followed by corresponding deletions. If GA subsumption is being used, each offspring is first tested to see if it is subsumed by any classifier in the current action set; if so, that offspring is not inserted in the population, and the subsuming parent's numerosity is increased.

$RUN\ GA([A],\ s,\ [P],\ \rho{:})$

1 if(time $t - \sum_{cl\in[A]} cl.ts \cdot cl.num\ /\ \sum_{cl\in[A]} cl.num > \theta_{GA}$)

2      $\bar{\varepsilon} \leftarrow 0$

3   for each classifier cl in $[A]$

4          $cl.ts \leftarrow$ actual time t

5          $\bar{\varepsilon} \leftarrow \bar{\varepsilon} + cl.\varepsilon$

6      $parent_1 \leftarrow$ SELECT OFFSPRING in $[A]$

7      $parent_2 \leftarrow$ SELECT OFFSPRING in $[A]$

8      $child_1 \leftarrow$ copy classifier $parent_1$

9      $child_2 \leftarrow$ copy classifier $parent_2$

10     $child_1.R \leftarrow child_2.R \leftarrow \rho$

11     $child_1.\varepsilon \leftarrow child_2.\varepsilon \leftarrow \bar{\varepsilon}$

12     $child_1.num \leftarrow child_2.num \leftarrow 1$

13     $child_1.exp \leftarrow child_2.exp \leftarrow 0$

14   if(RandomNumber $[0,1[< \chi$)

15   APPLY CROSSOVER on $child_1$ and $child_2$

16         $child_1.F \leftarrow child_2.F \leftarrow (parent_1.F + parent_2.F)/2$

17   for both children $child$

18         $child_F \leftarrow 0.1 * child_F$

19   APPLY MUTATION on $child$ according to s

20   if($doGASubsumption$)

21   if(DOES SUBSUME $parent_1, child$)

22                 $parent_1.num$++

23   else if(DOES SUBSUME $parent_2, child$)

24                 $parent_2.num$++

25   else

26   INSERT $child$ IN POPULATION

27   else

28   INSERT $child$ IN POPULATION

29   DELETE FROM POPULATION $[P]$

Proportionate Selection

Proportionate selection chooses a classifier for reproduction proportional to the fitness of the classifiers in set $[A]$. First, the sum of all the fitness values in the set $[A]$ is computed. Next, the roulette-wheel is spun. Finally, the classifier is chosen according to the roulette-wheel result.

SELECT OFFSPRING([A]):
```
1 fitnessSum ← 0
2 for each classifier cl in [A]
3     fitnessSum ← fitnessSum + cl.F
4 choicePoint ← RandomNumber [0,1[ * fitnessSum
5 fitnessSum ← 0
6 for each classifier cl in [A]
7     fitnessSum ← fitnessSum + cl.F
8     if(fitnessSum > choicePoint)
9         return cl
```

Tournament Selection

As shown in Chapter 5, tournament selection with tournament sizes proportional to the action set size proved to improve XCS's performance in all problem settings. The following algorithm describes the selection approach using tournament selection with an approximate tournament size of a fraction τ of the action set size.

SELECT OFFSPRING([A]):
```
1 clb ← null
2 while(clb = null)
3     maxf ← 0
4     for each classifier cl in [A]
5         if(cl.F/cl.num > maxf)
6             for each micro-classifier cl'
7                 if(RandomNumber [0,1) < τ)
8                     clb ← cl
9                     maxf ← cl.F/cl.num
10                    break
11 return clb
```

Crossover

The crossover procedure is similar to the standard crossover procedure in GAs. In the *APPLY CROSSOVER* procedure we show uniform crossover. The action part is not affected by crossover.

APPLY CROSSOVER(cl_1, cl_2):
```
1 for(i ← 0to Length of cl₁.C)
2     if(RandomNumber [0,1) < 0.5)
3         swap cl₁.C[i] and cl₂.C[i]
```

Mutation

While crossover does not affect the action, mutation takes place in both the condition and the action. A mutation in the condition flips the attribute to one of the other possibilities showing free mutation. Since in XCS most of the time is spent in processes that operate on the whole population such as matching and deletion, the algorithms used for mutation as well as crossover only slightly affect efficiency.

```
APPLY MUTATION(cl, σ):
 1 for(i ← 0 to length of cl.C)
 2     if(RandomNumber [0,1[ < μ)
 3         if(RandomNumber [0,1[ < 0.5)
 4             if(cl.C[i] = #)
 5                 cl.C[i] ←0
 6             else
 7                 cl.C[i] ← #
 8         else
 9             if(cl.C[i] = #)
10                 cl.C[i] ←1
11             else
12                 if(cl.C[i] = 0)
13                     cl.C[i] ←1
14                 else
15                     cl.C[i] ←0
16 if(RandomNumber [0,1[ < μ)
17     cl.A ← a randomly chosen other possible action
```

Insertion in and Deletion from the Population

This section covers processes that handle the insertion and deletion of classifiers in the current population $[P]$.

The *INSERT IN POPULATION* procedure searches for an identical classifier. If one exists, the latter's numerosity is incremented; if not, the new classifier is added to the population.

```
INSERT IN POPULATION(cl, [P]):
 1 for all c in [P]
 2     if(c is equal to cl in condition and action)
 3         c.num++
 4         return
 5 add cl to set [P]
```

Proportionate Deletion

The deletion procedure realizes two ideas at the same time: (1) It assures an approximately equal number of classifiers in each action set, or environmental 'niche'; (2) It removes low-fitness individuals from the population.

The deletion procedure chooses individuals (for deletion) by proportionate selection. Deletion is only triggered if the current population size is larger than N. If deletion is triggered, proportionate selection is applied based on the deletion vote. If the chosen classifier is a macro-classifier, its numerosity is decreased by one. Otherwise, the classifier is removed from the population.

DELETE FROM POPULATION([P]):
```
 1 if(∑_{c∈[P]} c.num < N)
 2    return
 3 voteSum ← 0
 4 for each classifier c in [P]
 5    voteSum ← voteSum + DELETION VOTE of c in [P]
 6 choicePoint ← RandomNumber [0,1) * voteSum
 7 voteSum ← 0
 8 for each classifier c in [P]
 9    voteSum ← voteSum + DELETION VOTE of c
10    if(voteSum > choicePoint)
11       if(c.num > 1)
12          c.num--
13       else
14          remove classifier c from set [P]
```

The Deletion Vote

As mentioned above, the deletion vote realizes niching as well as removal of the lowest fitness classifiers. The deletion vote of each classifier is based on the action set size estimate *as*. Moreover, if the classifier has sufficient experience and its fitness is significantly lower than the average fitness in the population, the vote is increased in inverse proportion to the fitness. In this calculation, since we are deleting one micro-classifier at a time, we need to use as fitness the (macro)classifier's fitness divided by its numerosity. The following *DELETION VOTE* procedure realizes all this.

DELETION VOTE(cl, [P]):
```
 1 vote ← cl.as * cl.num
 2 avFitnessInPopulation ← ∑_{c∈[P]} c.F / ∑_{c∈[P]} c.num
 3 if(cl.exp > θ_del and
       cl.F / cl.num < δ * avFitnessInPopulation)
 4    vote ← vote * avFitnessInPopulation / (cl.F / cl.num)
 5 return vote
```

Subsumption

GA subsumption checks an offspring classifier to see if its condition is logically subsumed by the condition of an accurate and sufficiently experienced action set member. If so, the offspring is not added to the population, but the subsumer's numerosity is incremented. If GA subsumption is applied, the sub-procedure *DOES SUBSUME* is called, which decides if the given classifier is subsumed by the parent.

For a classifier to subsume another classifier, it must first be sufficiently accurate and sufficiently experienced. This is tested by the *COULD SUBSUME* function. Then, if a classifier could be a subsumer, it must be tested to see if it has the same action and is really more general than the classifier that is to be subsumed. This is the case if the set of situations matched by the condition of the potentially subsumed classifier form a proper subset of the situations matched by the potential subsumer. The *IS MORE GENERAL* tests this.

DOES SUBSUME(cls, clt):
```
1 if(cls.A = clt.A)
2     if(cls COULD SUBSUME)
3         if(cls IS MORE GENERAL than clt)
4             return true
5 return false
```

COULD SUBSUME(cl):
```
1 if(cl.exp > θ_sub)
2     if(cl.ε < ε₀)
3         return true
4 return false
```

$COULD\ SUBSUME(cl):$
$$1\ \text{if}(cl.exp > \theta_{sub})$$
$$2\quad \text{if}(cl.\varepsilon < \varepsilon_0)$$
$$3\qquad \text{return } true$$
$$4\ \text{return } false$$

IS MORE GENERAL(clg, cls):
```
1 if (the number of # in clg.C ≤ the number of # in cls.C)
2     return false
3 for(i ← 0 to length of clg.C)
4     if(clg.C[i] ≠ # and clg.C[i] ≠ cls.C[i])
5         return false
6 return true
```

C

Boolean Function Problems

A short definition of all used binary classification (concept learning) problems is provided with augmenting examples. Problem properties are discussed. Some of the problems are augmented with a special reward scheme.

Multiplexer

Multiplexer problems have been shown to be challenging for other machine learning methods. De Jong and Spears (1991) showed the superiority of LCSs in comparison to other machine learning approaches such as C4.5 in the multiplexer problem. XCS has been applied to multiplexer problems beginning with its first publication (Wilson, 1995).

The multiplexer function is defined for binary strings of length $k + 2^k$. The output of the multiplexer function is determined by one of the 2^k value bits. The location is determined by the k address bits. For example, in the six multiplexer $f(100010) = 1$, $f(000111) = 0$, or $f(110101) = 1$. Any multiplexer can also be written in disjunctive normal form (DNF) in which there are 2^k conjunctions of length $k + 1$. The DNF of the 6-multiplexer is

$$6MP(x_1, x_2, x_3, x_4, x_5, x_6) = \neg x_1 \neg x_2 x_3 \vee \neg x_1 x_2 x_4 \vee x_1 \neg x_2 x_5 \vee x_1 x_2 x_6. \quad \text{(C.1)}$$

Using reward as feedback, a correct classification results in a payoff of 1000 while an incorrect classification results in a payoff of 0.

In terms of fitness guidance, the multiplexer problem provides slight fitness guidance, although not directly towards specifying the k position bits. Specifying any of the 2^k remaining bits gives the classifier a bias towards class zero or one. Thus, classifiers with more ones or zeros specified in the 2^k remaining bits have a lower error estimate on average. Specifying position bits (initially by chance) restricts the remaining relevant bits. These properties result in the fitness guidance in the multiplexer problem. The multiplexer problem needs $||[O]|| = 2^{k+2}$ accurate, maximally general classifiers in order to represent the whole problem accurately. With respect to problem difficulty, the multiplexer problem has a minimal order $k_m = 1$ and problem difficulty $k_d = k + 1$. The initial fitness guidance is actually somewhat misleading in that first the specialization of the value bits increases accuracy.

Interestingly, although there exists a complete, accurate, and maximally general non-overlapping problem solution, there are overlapping rules with equal generality and accuracy. For example, classifier 0#00##→0 is accurate and maximally general but it overlaps with classifiers 000###→0 and 01#0##→0. Due to the space distribution, the latter two classifiers are expected to erase the former. However, due to continuous search, the former can be expected to continuously reappear.

Layered Multiplexer

A layered reward scheme for the multiplexer problem was introduced in Wilson (1995), intending to show that XCS is able to handle more than two reward levels (particularly relevant for multistep problems, in which reward is discounted and propagated). Later (as shown in Chapter 8), it was shown that the layered reward scheme is actually easier for XCS since the scheme provides stronger fitness guidance (Butz, Kovacs, Lanzi, & Wilson, 2004).

Reward of a specific instance-classification case is determined by

$$R(S, A) = (\text{value of } k \text{ position bits} + \text{return value }) \cdot 100 + \text{correctness} \cdot 300 \tag{C.2}$$

This function assures that the more position bits are specified, the less different the resulting reward values can be. Thus, XCS successively learns to have all position bits specified beginning with the left-most one. Similar to the 'normal' multiplexer problem above, the layered multiplexer needs $|[O]| = 2^{k+2}$ accurate, maximally general classifiers in order to represent the whole problem accurately.

Also with respect to problem difficulty, the layered multiplexer is equal to the normal multiplexer. However, due to the layered reward scheme, fitness guidance immediately propagates position attributes and overlapping classifiers are not as prominent as in the normal multiplexer problem.

xy-Biased Multiplexer

The xy-biased multiplexer was originally designed to investigate fitness guidance. The problem is designed to exhibit more direct fitness guidance.

The problem is composed of x biased multiplexer problems. A *biased multiplexer* is a Boolean multiplexer whose output is biased towards outcome zero or one. We can distinguish a *zero-biased multiplexer*, in which the zero output is more likely to be correct, and a *one-biased multiplexer*, in which the one output is more likely to be correct. A biased multiplexer, is defined over $l = k + (2^k - 1)$ bits; as in the Boolean multiplexer the first k bits represent an address which indexes the remaining $2^k - 1$ bits. In a biased multiplexer, it is not possible to address one of the configurations: k address bits would address 2^k bits, but in this case only $2^k - 1$ bits are available. The missing configuration results in the bias. A *one-biased multiplexer* always returns one when the k address bits are all 1, i.e., the output one is correct for one configuration more than would be in the case in a Boolean multiplexer with k address bits. For instance, in a *one-biased multiplexer* with two address bits the condition 11### always corresponds to the output 1. A *zero-biased multiplexer* always returns zero when the k address bits are all 0s.

A set of *biased multiplexers* can be used to build an xy-biased multiplexer. The problem uses the x reference bits to refer to one of the 2^x biased multiplexers. y refers to the size of each one of the biased multiplexers involved. The first half of the 2^x biased multiplexers are zero-biased and the second half are one-biased. Overall, a problem instance is of length $l = x + 2^x \cdot (y + 2^y - 1)$ bits.

As an example, let us consider the *11-biased multiplexer* (*11bMP*). The first bit of the multiplexer, x_0, refers to one of the two biased multiplexers, which consist of two bits; the first biased multiplexer is *zero biased*, the second biased multiplexer is *one biased*. The *11-biased multiplexer* for input 00111 outputs 0 because the first zero refers to the (first) zero biased multiplexer and the second zero determines output zero; input 01011 has output 0 as well; input 10010 has output 1 since the one-biased multiplexer is now referenced; input 10000 would be output 0. The *11-biased multiplexer* can be written in DNF as:

$$11 - bMP(x_1, x_2, x_3, x_4, x_5) = \neg x_1 x_2 x_3 \vee x_1 x_4 \vee x_1 \neg x_4 x_5. \qquad \text{(C.3)}$$

Again, using reward as feedback, a reward of 1000 indicates a correct classification and a reward of 0 an incorrect one.

To represent a y-biased multiplexer accurately, XCS needs to evolve $\cdot 2^{y+2} - 2$ classifiers. Thus, the complete, accurate, and maximally general problem representation in the xy-biased multiplexer requires $|[O]| = 2^{x+y+2} - 2^{x+1}$ classifiers. Due to the bias in the outcome, XCS can be expected to detect the x reference bits quickly. Once the relevancy of one biased-multiplexer is specified, first the address bits and then the value bits are expected to be correctly specified.

Hidden Parity Function

Kovacs and Kerber (2001) triggered the notion of a hidden parity function to show the dependence of the problem difficulty on the necessary number of accurate, maximally general classifiers $[O]$ in XCS. Hidden parity functions are the extreme case of a BB of size k. It is basically identical to an XOR function with additional, irrelevant input bits. In the hidden parity function only k bits are relevant in a string of length l. The number of ones modulo two determines the class of the input string. For example, in a $k = 3$, $l = 6$ hidden parity problem (in which the first k bits are assigned to be the relevant ones) string 110000 as well as 011000 would be in class zero while string 010000 as well as string 111000 would be class one. In DNF, the hidden parity problem with $k = 3$ and $l = 6$ can be written as:

$$HP3(x_1, x_2, x_3, x_4, x_5, x_6) = x_1 x_2 x_3 \lor \neg x_1 \neg x_2 x_3 \lor \neg x_1 x_2 \neg x_3 \lor x_1 \neg x_2 \neg x_3$$
(C.4)

The reward scheme is again 1000 for a correct classification and zero otherwise.

The hidden parity problem is the most difficult problem with respect to problem difficulty since the minimal order of difficulty is equal to the overall problem difficulty ($k_m = k_d = k$). Thus, XCS needs to have classifiers with all k relevant attributes specified, or needs to generate them by chance due to mutation as described in Chapter 6. The complete problem solution requires the specialization of all k attributes and thus the optimal solution is of size $||[O]|| = 2^{k+1}$.

Count Ones

The count ones problem is similar to the hidden parity function in that only k positions in a string of length l are relevant. However, it is very different in that the schemata with minimal order that provide fitness guidance are of order one ($k_m = 1$). Any specialization of only ones or only zeros in the k relevant attributes makes a classifier more accurate.

The class in the count ones problem is defined by the number of ones in the k relevant bits. If this number is greater than half k, the class is one, otherwise the class is zero. For example, considering a count ones problem of length $l = 5$ with $k = 3$ relevant attributes, problem instances 11100 as well as 01111 belong to class one whereas 01011 or 00011 belong to class zero. In DNF, the problem can be written as:

$$CO3(x_1, x_2, x_3, x_4, x_5) = x_1 x_2 \lor x_1 x_3 \lor x_2 x_3.$$
(C.5)

We again use the 1000/0 reward scheme in this problem. Due to the counting approach, even a specialization of only one attribute of the k relevant attributes biases the probability of correctness and consequently decreases the prediction error estimate. Thus, this problem provides a very strong fitness guidance and $k_m = 1$. The order of difficulty equals $\lceil \frac{k}{2} \rceil$ since more than half of the features need to be one to determine class one. The number of accurate, maximally general classifiers is $[O] = 4\binom{k}{\lceil \frac{k}{2} \rceil}$. Note that the optimal set of accurate, maximally general classifier is strongly overlapping.

Layered Count Ones

The layered count ones problem is identical to the count ones problem except for the reward scheme used. The received reward now depends on the number of ones in the string. That is, a reward of $1000 \cdot \sum_{i=1}^{k} value[i]$ is provided if the classification was correct, and 1000 minus this value otherwise. This scheme causes any specialization in the k position bits to result in a lower deviation of reward outcomes and thus in a lower reward error estimate. Thus, like the count ones problem, the layered count ones problem provides a very strong fitness guidance, and the evolutionary process can be expected to stay rather independent of the string length. However, note that there are more classes to distinguish than in the layered count ones problem. Since any change in one of the relevant bits changes the resulting reward, all k bits need to be specified in any classifier that specifies the input string accurately. Thus, the number of accurate, maximally general classifiers $||[O]|| = 2^{k+1}$. Unlike in the count ones problem, the accurate, maximally general classifiers do not overlap.

Carry Problem

The carry problem adds the additional difficulty of strongly differently sized problem niches to the spectrum of problems. That is, different solution sub-spaces require different numbers of specialized attributes.

In the carry problem, essentially, two binary numbers of length k are added together. If the addition yields a carry (that is, if the addition results in a number greater than expressible with the number of bits used), then the class of the problem instance is one, and zero otherwise. For example, the 3,3-carry problem adds two binary numbers of length three so that problem instance 001011 yields value 100 (class zero) or 101010 yields value 111 (class zero). One the other hand, 100100 yields value 1000 and thus class one, as does 111001. The carry problem can be written in DNF-form as follows:

$$3, 3 - Carry(x_1, x_2, x_3, x_4, x_5, x_6) =$$
$$x_1 x_4 \vee x_1 x_2 x_5 \vee x_2 x_4 x_5 \vee x_1 x_2 x_3 x_6 \vee x_1 x_3 x_5 x_6 \vee x_2 x_3 x_4 x_6 \vee x_3 x_4 x_5 x_6.$$
$$(C.6)$$

Essentially, in all carry problems the lower order accurate subsolution is of order two (specifying either both left bits as zero or as one) and the maximal order accurate subsolution is of order $k + 1$, specifying that a carry happens due to the propagation of the carry starting right at the back of the num-ber (see the latter conjunctions in the DNF-form). In effect, the size of the optimal problem solution is $\sum_{i=1}^{k} 2^{i-1} = 2^k - 1$ to represent class one and $2^k + \sum_{i=1}^{k-1} 2^{i-1} = 2^k + 2^{k-1} - 1$ to represent class zero. Thus, the optimal solu-tion representation in XCS comprises $||[O]|| = 2(2^k - 1 + 2^k + 2^{k-1} - 1) = 5 \cdot 2^k - 4$ classifiers.

Similar to the count ones problems, the carry problem provides helpful fitness guidance in that the specialization of only ones or only zeros increases accuracy (decreasing entropy in the class distribution). Thus, the minimal order of difficulty $k_m = 1$ and the order of difficulty $k_d = k + 1$. Due to the overlapping nature and the unequally sized niches in the problem, XCS

Fig. C.1. Illustration of an exemplar hierarchical 3-parity, 6-multiplexer problem evaluation

requires larger population sizes to ensure niche support as evaluated in Chapter 6.

Hierarchically Composed Problems

We introduced general hierarchal problems in Chapter 7, showing that they require effective BB processing in XCS. The hierarchical problems are designed in a two-level hierarchy in which the lower-level is evaluated by one set of Boolean functions and the output of the lower-level is then fed as input to the higher level.

For example, we can combine parity problems on the lower level with a multiplexer problem on the higher level. The evaluation takes a problem instance and evaluates chunks of bits using the lower level function. The result of the lower level function results in a shorter bit string which is then evaluated using the higher level function. For example, consider the hierarchical 2-parity, 3-multiplexer problem. The problem is six bits long. On the lower level, blocks of two bits are evaluated by the parity function, that is, if there are an even number of ones, the result will be zero and one otherwise. The result is a string of three bits that is then evaluated by the 3-multiplexer function. The result is the class of the problem instance. The hierarchical 2-parity, 3-multiplexer problem can be written in DNF as follows:

$$2\text{-PA},3\text{-MP}(x_1, x_2, x_3, x_4, x_5, x_6) = x_1 x_2 \neg x_3 x_4 \vee x_1 x_2 x_3 \neg x_4 \vee$$
$$\neg x_1 \neg x_2 \neg x_3 x_4 \vee \neg x_1 \neg x_2 x_3 \neg x_4 \vee \neg x_1 x_2 \neg x_5 x_6 \vee$$
$$\neg x_1 x_2 x_5 \neg x_6 \vee x_1 \neg x_2 \neg x_5 x_6 \vee x_1 \neg x_2 x_5 \neg x_6. \tag{C.7}$$

Figure C.1 shows the structure of the hierarchical 3-parity-6-multiplexer problem. The figure shows the tight coding of the problem: All 3-parity chunks are coded in one block. Certainly, the property of tight coding may not be given in a natural problem and thus our learning algorithm should not rely on this property.

With respect to problem difficulty, generally the previously determined sizes carry over only that each problem feature is now represented by a lower-level BB of features defined by the lower level functions. In the k-parity-k'-multiplexer combination, effectively the optimal population is of size $||O|| = 2(2^{k(k'+1)})$. The minimal order of difficulty $k_m = k$ and the problem order of difficulty $k_d = k(k' + 1)$.

References

[Aha, Kibler, & Albert (1991] Aha, D. W., Kibler, D., & Albert, M. K. (1991). Instance-based learning algorithms. *Machine Learning*, *6*, 37–66.

[Bacardit & Butz (2004] Bacardit, J., & Butz, M. V. (2004). *Data mining in learning classifier systems: Comparing XCS with GAssist* (IlliGAL report 2004030). University of Illinois at Urbana-Champaign: Illinois Genetic Algorithms Laboratory.

[Bäck & Schwefel (1995] Bäck, T., & Schwefel, H.-P. (1995). Evolution strategies I: Variants and their computational implementation. In Winter, G., Périaux, J., Gal'an, M., & Cuesta, P. (Eds.), *Genetic algorithms in engineering and computer science* (Chapter 7, pp. 111–126). Chichester: John Wiley & Sons.

[Baird (1995] Baird, L. C. (1995). Residual algorithms: Reinforcement learning with function approximation. *Proceedings of the Twelfth International Conference on Machine Learning*, 30–37.

[Baird (1999] Baird, L. C. (1999). *Reinforcement learning through gradient descent*. Doctoral dissertation, School of Computer Science. Carnegie Mellon University, Pittsburgh, PA 15213.

[Baker (1985] Baker, J. (1985). Adaptive selection methods for genetic algorithms. In Grefenstette, J. J. (Ed.), *Proceedings of the international conference on genetic algorithms and their applications* (pp. 14–21). Hillsdale, NJ: Lawrence Erlbaum Associates.

[Barto & Mahadevan (2003] Barto, A. G., & Mahadevan, S. (2003). Recent advances in hierarchical reinforcement learning. *Discrete Event Dynamic Systems*, *13*, 341–379.

[Bellman (1957] Bellman, R. (1957). *Dynamic programming*. Princeton, NY: Princeton University Press.

[Bernadó, Llorà, & Garrell (2002] Bernadó, E., Llorà, X., & Garrell, J. M. (2002). XCS and GALE: A comparative study of two learning classifier systems and six other learning algorithms on classification tasks. In Lanzi, P. L., Stolzmann, W., & Wilson, S. W. (Eds.), *Advances in*

Learning Classifier Systems (LNAI 2321) (pp. 115–132). Berlin Heidelberg: Springer-Verlag.

[Bernadó-Mansilla & Garrell-Guiu (2003] Bernadó-Mansilla, E., & Garrell-Guiu, J. M. (2003). Accuracy-based learning classifier systems: Models, analysis, and applications to classification tasks. *Evolutionary Computation*, *11*, 209–238.

[Booker (1982] Booker, L. (1982). *Improving behavior as an adaptation to the task environment*. Doctoral dissertation, Department of Computer and Communication Sciences, University of Michigan, Ann Arbor.

[Booker (1993] Booker, L. B. (1993). Recombination distributions for genetic algorithms. *Foundations of Genetic Algorithms*, *2*, 29–44.

[Booker, Goldberg, & Holland (1989] Booker, L. B., Goldberg, D. E., & Holland, J. H. (1989). Classifier systems and genetic algorithms. *Artificial Intelligence*, *40*, 235–282.

[Bridges & Goldberg (1987] Bridges, C. L., & Goldberg, D. E. (1987). An analysis of reproduction and crossover in a binary-coded genetic algorithm. *Proceedings of the Second International Conference on Genetic Algorithms*, 9–13.

[Bull (2003] Bull, L. (2003). *Investigating fitness sharing in a simple payoff-based learning classifier system* (Technical Report UWELCSG03-009). Bristol, UK: University of Western England, Learning Classifier System Group.

[Bull & Hurst (2002] Bull, L., & Hurst, J. (2002). ZCS redux. *Evolutionary Computation*, *10*(2), 185–205.

[Buntine (1991] Buntine, W. L. (1991). Theory refinement of Bayesian networks. *Proceedings of the Uncertainty in Artificial Intelligence (UAI-91)*, 52–60.

[Butz (2002] Butz, M. V. (2002). *Anticipatory learning classifier systems*. Boston, MA: Kluwer Academic Publishers.

[Butz (2004a] Butz, M. V. (2004a). Anticipation for learning, cognition, and education. *On the Horizon* (12), 111–116.

[Butz (2004b] Butz, M. V. (2004b). *COSEL: A cognitive sequence learning architecture* (IlliGAL report 2004021). University of Illinois at Urbana-Champaign: Illinois Genetic Algorithms Laboratory.

[Butz, Goldberg, & Lanzi (2004] Butz, M. V., Goldberg, D. E., & Lanzi, P. L. (2004). Bounding learning time in XCS. *Proceedings of the Sixth Genetic and Evolutionary Computation Conference (GECCO-2004): Part II*, 739–750.

[Butz, Goldberg, Lanzi, & Sastry (2004] Butz, M. V., Goldberg, D. E., Lanzi, P. L., & Sastry, K. (2004). *Bounding the population size to ensure niche support in XCS* (IlliGAL report 2004033). University of Illinois at Urbana-Champaign: Illinois Genetic Algorithms Laboratory.

[Butz, Goldberg, & Stolzmann (2002] Butz, M. V., Goldberg, D. E., & Stolzmann, W. (2002). The anticipatory classifier system and genetic generalization. *Natural Computing*, *1*, 427–467.

[Butz, Goldberg, & Tharakunnel (2003] Butz, M. V., Goldberg, D. E., & Tharakunnel, K. (2003). Analysis and improvement of fitness exploitation in XCS: Bounding models, tournament selection, and bilateral accuracy. *Evolutionary Computation*, *11*, 239–277.

[Butz, Kovacs, Lanzi, & Wilson (2001] Butz, M. V., Kovacs, T., Lanzi, P. L., & Wilson, S. W. (2001). How XCS evolves accurate classifiers. *Proceedings of the Third Genetic and Evolutionary Computation Conference (GECCO-2001)*, 927–934.

[Butz, Kovacs, Lanzi, & Wilson (2004] Butz, M. V., Kovacs, T., Lanzi, P. L., & Wilson, S. W. (2004). Toward a theory of generalization and learning in XCS. *IEEE Transactions on Evolutionary Computation*, *8*, 28– 46.

[Butz, Sigaud, & Gérard (2003] Butz, M. V., Sigaud, O., & Gérard, P. (2003). Internal models and anticipations in adaptive learning systems. In Butz, M. V., Sigaud, O., & Gérard, P. (Eds.), *Anticipatory Behavior in Adaptive Learning Systems: Foundations, Theories, and Systems* (pp. 86–109). Berlin Heidelberg: Springer-Verlag.

[Butz, Sigaud, & Swarup (2004] Butz, M. V., Sigaud, O., & Swarup, S. (2004). Anticipatory behavior in adaptive learning systems, ABiALS 2004 workshop proceedings.

[Butz, Swarup, & Goldberg (2004] Butz, M. V., Swarup, S., & Goldberg, D. E. (2004). *Effective online detection of task-independent landmarks* (IlliGAL report 2004002). University of Illinois at Urbana-Champaign: Illinois Genetic Algorithms Laboratory.

[Butz & Wilson (2001] Butz, M. V., & Wilson, S. W. (2001). An algorithmic description of XCS. In Lanzi, P. L., Stolzmann, W., & Wilson, S. W. (Eds.), *Advances in learning classifier systems: Third international workshop, IWLCS 2000 (LNAI 1996)* (pp. 253–272). Berlin Heidelberg: Springer-Verlag.

[Butz & Wilson (2002] Butz, M. V., & Wilson, S. W. (2002). An algorithmic description of XCS. *Soft Computing*, *6*, 144–153.

[Chickering, Heckerman, & Meek (1997] Chickering, D. M., Heckerman, D., & Meek, C. (1997). *A Bayesian approach to learning Bayesian networks with local structure* (Technical Report MSR-TR-97-07). Redmond, WA: Microsoft Research.

[Cooper & Herskovits (1992] Cooper, G. F., & Herskovits, E. H. (1992). A Bayesian method for the induction of probabilistic networks from data. *Machine Learning*, *9*, 309–347.

[Darwin (1859] Darwin, C. (1968 (orig. 1859)). *The origin of species by means of natural selection*. Penguin Books.

[De Jong (1975] De Jong, K. A. (1975). *An analysis of the behavior of a class of genetic adaptive systems*. Doctoral dissertation, University of Michigan, Ann Arbor. University Microfilms No. 76-9381.

[De Jong & Spears (1991] De Jong, K. A., & Spears, W. M. (1991). Learning concept classification rules using genetic algorithms. *IJCAI-91 Proceed-*

ings of the Twelfth International Conference on Artificial Intelligence, 651–656.

[DeJong, Spears, & Gordon (1993] DeJong, K. A., Spears, W. M., & Gordon, D. F. (1993). Using genetic algorithms for concept learning. *Machine Learning*, *13*(2/3), 161–188.

[Dietterich (1997] Dietterich, T. G. (1997). Machine learning research: Four current directions. *AI Magazine*, *18*(4), 97–136.

[Dietterich (2000] Dietterich, T. G. (2000). Hierarchical reinforcement learning with the MAXQ value function decomposition. *Journal of Artificial Intelligence Research*, *13*, 227–303.

[Dittenbach, Merkl, & Rauber (2000] Dittenbach, M., Merkl, D., & Rauber, A. (2000). The growing hierarchical self-organizing map. *Proceedings of the International Joint Conference on Neural Networks*, *VI*, 15–19.

[Drummond (2002] Drummond, C. (2002). Accelerating reinforcement learning by composing solutions of automatically identified subtasks. *Journal of Artificial Intelligence Research*, *16*, 59–104.

[Duda, Hart, & Stork (2001] Duda, R. O., Hart, P. E., & Stork, D. G. (2001). *Pattern classification* (2 ed.). New York, NY: John Wiley & Sons.

[Feng & Ratnam (2000] Feng, A. S., & Ratnam, R. (2000). Neural basis of hearing in real-world situations. *Annual Review of Psychology*, *51*, 699–725.

[Frank & Witten (1998] Frank, E., & Witten, I. H. (1998). Generating accurate rule sets without global optimization. *Proceedings of the Fifteenth International Conference on Machine Learning*, 144–151.

[Friedman & Goldszmidt (1999] Friedman, N., & Goldszmidt, M. (1999). Learning Bayesian networks with local structure. In Jordan, M. I. (Ed.), *Graphical Models* (pp. 421–459). Cambridge, MA: MIT Press.

[Gelb, Kasper, Nash, Price, & Sutherland (1974] Gelb, A., Kasper, J. F., Nash, R. A., Price, C. F., & Sutherland, A. A. (Eds.) (1974). *Applied optimal estimation*. Cambridge, MA: MIT Press.

[Gérard & Sigaud (2001a] Gérard, P., & Sigaud, O. (2001a). Adding a generalization mechanism to YACS. *Proceedings of the Third Genetic and Evolutionary Computation Conference (GECCO-2001)*, 951–957.

[Gérard & Sigaud (2001b] Gérard, P., & Sigaud, O. (2001b). YACS: Combining dynamic programming with generalization in classifier systems. In Lanzi, P. L., Stolzmann, W., & Wilson, S. W. (Eds.), *Advances in learning classifier systems: Third international workshop, IWLCS 2000 (LNAI 1996)* (pp. 52–69). Berlin Heidelberg: Springer-Verlag.

[Gérard & Sigaud (2003] Gérard, P., & Sigaud, O. (2003). Designing efficient exploration with MACS: Modules and function approximation. *Proceedings of the Fifth Genetic and Evolutionary Computation Conference (GECCO-2003)*, 1882–1893.

[Gibson (1979] Gibson, J. J. (1979). *The ecological approach to visual perception*. Mahwah, NJ: Lawrence Erlbaum Associates.

[Giese & Poggio (2003] Giese, M. A., & Poggio, T. (2003). Neural mechanisms for the recogniton of biological movements. *Nature Reviews Neuroscience*, *4*, 179–192.

[Goldberg & Richardson (1987] Goldberg, D., & Richardson, J. (1987). Genetic algorithms with sharing for multimodal function optimization. *Proceedings of the Second International Conference on Genetic Algorithms*, 41–49.

[Goldberg (1983] Goldberg, D. E. (1983). Computer-aided gas pipeline operation using genetic algorithms and rule learning. *Dissertation Abstracts International*, *44*(10), 3174B. Doctoral dissertation, University of Michigan.

[Goldberg (1989] Goldberg, D. E. (1989). *Genetic algorithms in search, optimization and machine learning*. Reading, MA: Addison-Wesley.

[Goldberg (1990] Goldberg, D. E. (1990). Probability matching, the magnitude of reinforcement, and classifier system bidding. *Machine Learning*, *5*(4), 407–425.

[Goldberg (1991] Goldberg, D. E. (1991). Genetic algorithms as a computational theory of conceptual design. In Rzevski, G., & Adey, R. A. (Eds.), *Applications of Artificial Intelligence in Engineering VI* (pp. 3–16). Boston and New York: Computational Mechanics Publications and Elsevier Applied Science.

[Goldberg (1999] Goldberg, D. E. (1999). The race, the hurdle and the sweet spot: Lessons from genetic algorithms for the automation of innovation and creativity. In Bentley, P. (Ed.), *Evolutionary design by computers* (pp. 105–118). San Francisco, CA: Morgan Kaufmann.

[Goldberg (2002] Goldberg, D. E. (2002). *The design of innovation: Lessons from and for competent genetic algorithms*. Boston, MA: Kluwer Academic Publishers.

[Goldberg & Deb (1991] Goldberg, D. E., & Deb, K. (1991). A comparative analysis of selection schemes used in genetic algorithms. *Foundations of Genetic Algorithms*, 69–93.

[Goldberg, Deb, & Clark (1992] Goldberg, D. E., Deb, K., & Clark, J. H. (1992). Genetic algorithms, noise, and the sizing of populations. *Complex Systems*, *6*, 333–362.

[Goldberg, Deb, & Thierens (1993] Goldberg, D. E., Deb, K., & Thierens, D. (1993). Toward a better understanding of mixing in genetic algorithms. *Journal of the Society of Instrument and Control Engineers*, *32*(1), 10–16.

[Goldberg & Sastry (2001] Goldberg, D. E., & Sastry, K. (2001). A practical schema theorem for genetic algorithm design and tuning. *Proceedings of the Third Genetic and Evolutionary Computation Conference (GECCO-2001)*, 328–335.

[Grossberg (1976a] Grossberg, S. (1976a). Adaptive pattern classification and universal recoding, I: Parallel development and coding of neural feature detectors. *Biological Cybernetics*, *23*, 121–134.

[Grossberg (1976b] Grossberg, S. (1976b). Adaptive pattern classification and universal recoding, II: Feedback, expectation, olfaction, and illusions. *Biological Cybernetics*, *23*, 197–202.

[Harik (1994] Harik, G. (1994). *Finding multiple solutions in problems of bounded difficulty* (IlliGAL report 94002). University of Illinois at Urbana-Champaign: Illinois Genetic Algorithms Laboratory.

[Harik (1999] Harik, G. (1999). *Linkage learning via probabilistic modeling in the ECGA* (IlliGAL report 99010). University of Illinois at Urbana-Champaign: Illinois Genetic Algorithms Laboratory.

[Harik, Cantú-Paz, Goldberg, & Miller (1997] Harik, G., Cantú-Paz, E., Goldberg, D. E., & Miller, B. (1997). The gambler's ruin problem, genetic algorithms, and the sizing of populations. *Proceedings of the Fourth International Conference on Evolutionary Computation*, 7–12.

[Hassoun (1995] Hassoun, M. H. (1995). *Fundamentals of artificial neural networks*. Cambridge, MA: MIT Press.

[Haykin (1999] Haykin, S. (1999). *Neural networks: A comprehensive foundation*. Upper Saddle River, NJ: Prentice Hall. 2nd edition.

[Haykin (2001] Haykin, S. (Ed.) (2001). *Kalman filtering and neural networks*. New York: John Wiley & Sons.

[Heckerman, Geiger, & Chickering (1994] Heckerman, D., Geiger, D., & Chickering, D. M. (1994). *Learning Bayesian networks: The combination of knowledge and statistical data* (Technical Report MSR-TR-94-09). Redmond, WA: Microsoft Research.

[Henrion (1988] Henrion, M. (1988). Propagating uncertainty in Bayesian networks by probabilistic logic sampling. In Lemmer, J. F., & Kanal, L. N. (Eds.), *Uncertainty in Artificial Intelligence* (pp. 149–163). Amsterdam, London, New York: Elsevier.

[Hoffmann (1993] Hoffmann, J. (1993). *Vorhersage und Erkenntnis: Die Funktion von Antizipationen in der menschlichen Verhaltenssteuerung und Wahrnehmung. [Anticipation and cognition: The function of anticipations in human behavioral control and perception.]*. Göttingen, Germany: Hogrefe.

[Hoffmann (2003] Hoffmann, J. (2003). Anticipatory behavioral control. In Butz, M. V., Sigaud, O., & Gérard, P. (Eds.), *Anticipatory Behavior in Adaptive Learning Systems: Foundations, Theories, and Systems* (pp. 44–65). Berlin Heidelberg: Springer-Verlag.

[Holland (1971] Holland, J. H. (1971). Processing and processors for schemata. In Jacks, E. L. (Ed.), *Associative Information Techniques* (pp. 127–146). New York: American Elsevier.

[Holland (1975] Holland, J. H. (1975). *Adaptation in natural and artificial systems*. Ann Arbor, MI: University of Michigan Press. second edition, 1992.

[Holland (1976] Holland, J. H. (1976). Adaptation. In Rosen, R., & Snell, F. (Eds.), *Progress in theoretical biology*, Volume 4 (pp. 263–293). New York: Academic Press.

[Holland (1977] Holland, J. H. (1977). *A cognitive system with powers of generalization and adaptation.* Unpublished manuscript.

[Holland (1985] Holland, J. H. (1985). Properties of the bucket brigade algorithm. *Proceedings of an International Conference on Genetic Algorithms and their Applications*, 1–7.

[Holland (1995] Holland, J. H. (1995). *Hidden order: How adaptation builds complexity.* Perseus Book.

[Holland & Reitman (1978] Holland, J. H., & Reitman, J. S. (1978). Cognitive systems based on adaptive algorithms. In Waterman, D. A., & Hayes-Roth, F. (Eds.), *Pattern directed inference systems* (pp. 313–329). New York: Academic Press.

[Holte (1993] Holte, R. C. (1993). Very simple classification rules perform well on most commonly used datasets. *Machine Learning*, *11*, 63–90.

[Hommel, Müsseler, Aschersleben, & Prinz (2001] Hommel, B., Müsseler, J., Aschersleben, G., & Prinz, W. (2001). The theory of event coding (TEC): A framework for perception and action planning. *Behavioral and Brain Sciences*, *24*, 849–878.

[Horn (1993] Horn, J. (1993). Finite Markov chain analysis of genetic algorithms with niching. *Proceedings of the Fifth International Conference on Genetic Algorithms*, 110–117.

[Horn, Goldberg, & Deb (1994] Horn, J., Goldberg, D. E., & Deb, K. (1994). Implicit niching in a learning classifier system: Nature's way. *Evolutionary Computation*, *2*(1), 37–66.

[Howard & Matheson (1981] Howard, R. A., & Matheson, J. E. (1981). Influence diagrams. In Howard, R. A., & Matheson, J. E. (Eds.), *Readings on the principles and applications of decision analysis*, Volume II (pp. 721–762). Menlo Park, CA: Strategic Decisions Group.

[Jaeger (2000] Jaeger, H. (2000). Observable operator models for discrete stochastic time series. *Neural Computation*, *12*, 1371–1398.

[Jaeger (2004] Jaeger, H. (2004). Online learning algorithms for observable operator models.

[James & Singh (2004] James, M. R., & Singh, S. (2004). Learning and discovery of predictive state representations in dynamical systems with reset. *Proceedings of the Twenty-First International Conference on Machine Learning (ICML-2004)*, 417–424.

[John & Langley (1995] John, G. H., & Langley, P. (1995). Estimating continuous distributions in Bayesian classifiers. *11th Conference on Uncertainty in Artificial Intelligence*, 338–345.

[Kaelbling, Littman, & Moore (1996] Kaelbling, L. P., Littman, M. L., & Moore, A. W. (1996). Reinforcement learning: A survey. *Journal of Artificial Intelligence Research*, *4*, 237–285.

[Kalman (1960] Kalman, R. E. (1960). A new approach to linear filtering and prediction problems. *Transactions of the ASME-Journal of Basic Engineering*, *82*(Series D), 35–45.

[Kersting, Van Otterlo, & De Raedt (2004] Kersting, K., Van Otterlo, M., & De Raedt, L. (2004). Bellman goes relational. *Proceedings of the Twenty-First International Conference on Machine Learning (ICML-2004)*, 465–472.

[Kovacs (1996] Kovacs, T. (1996). *Evolving optimal populations with XCS classifier systems*. Master's thesis, School of Computer Science, University of Birmingham, Birmingham, U.K.

[Kovacs (1997] Kovacs, T. (1997). XCS classifier system reliably evolves accurate, complete, and minimal representations for boolean functions. In Roy, Chawdhry, & Pant (Eds.), *Soft computing in engineering design and manufacturing* (pp. 59–68). Springer-Verlag, London.

[Kovacs (1999] Kovacs, T. (1999). Deletion schemes for classifier systems. *Proceedings of the Genetic and Evolutionary Computation Conference (GECCO-99)*, 329–336.

[Kovacs (2000] Kovacs, T. (2000). Strength or Accuracy? Fitness calculation in learning classifier systems. In Lanzi, P. L., Stolzmann, W., & Wilson, S. W. (Eds.), *Learning classifier systems: From foundations to applications (LNAI 1813)* (pp. 143–160). Berlin Heidelberg: Springer-Verlag.

[Kovacs (2001] Kovacs, T. (2001). Towards a theory of strong overgeneral classifiers. *Foundations of Genetic Algorithms 6*, 165–184.

[Kovacs (2003] Kovacs, T. (2003). *Strength or accuracy: Credit assignment in learning classifier systems*. Berlin Heidelberg: Springer-Verlag.

[Kovacs & Kerber (2001] Kovacs, T., & Kerber, M. (2001). What makes a problem hard for XCS? In Lanzi, P. L., Stolzmann, W., & Wilson, S. W. (Eds.), *Advances in learning classifier systems: Third international workshop, IWLCS 2000 (LNAI 1996)* (pp. 80–99). Berlin Heidelberg: Springer-Verlag.

[Kovacs & Yang (2004] Kovacs, T., & Yang, C. (2004). An initial study of a model-based XCS. http://www.psychologie.uni-wuerzburg.de/IWLCS/.

[Kunde, Koch, & Hoffmann (2004] Kunde, W., Koch, I., & Hoffmann, J. (2004). Anticipated action effects affect the selection, initiation, and execution of actions. *The Quarterly Journal of Experimental Psychology. Section A: Human Experimental Psychology, 57*, 87–106.

[Lanzi (1997] Lanzi, P. L. (1997). A study of the generalization capabilities of XCS. *Proceedings of the Seventh International Conference on Genetic Algorithm*, 418–425.

[Lanzi (1999a] Lanzi, P. L. (1999a). An analysis of generalization in the XCS classifier system. *Evolutionary Computation, 7*(2), 125–149.

[Lanzi (1999b] Lanzi, P. L. (1999b). Extending the Representation of Classifier Conditions Part II: From Messy Coding to S-Expressions. *Proceedings of the Genetic and Evolutionary Computation Conference (GECCO-99)*, 345–352.

[Lanzi (1999c] Lanzi, P. L. (1999c). An extension to the XCS classifier system for stochastic environments. *Proceedings of the Genetic and Evolutionary Computation Conference (GECCO-99)*, 353–360.

[Lanzi (2000] Lanzi, P. L. (2000). Adaptive agents with reinforcement learning and internal memory. *From Animals to Animats 6: Proceedings of the Sixth International Conference on Simulation of Adaptive Behavior*, 333–342.

[Lanzi (2002] Lanzi, P. L. (2002). Learning classifier systems from a reinforcement learning perspective. *Soft Computing: A Fusion of Foundations, Methodologies and Applications*, *6*, 162–170.

[Lanzi, Loiacono, Wilson, & Goldberg (2005] Lanzi, P. L., Loiacono, D., Wilson, S. W., & Goldberg, D. E. (2005). Extending XCSF beyond linear approximation.

[Lanzi & Wilson (2000] Lanzi, P. L., & Wilson, S. W. (2000). Toward optimal classifier system performance in non-Markov environments. *Evolutionary Computation*, *8*(4), 393–418.

[Larrañaga (2002] Larrañaga, P. (2002). A review on estimation of distribution algorithms. In Larrañaga, P., & Lozano, J. A. (Eds.), *Estimation of Distribution Algorithms* (Chapter 3, pp. 57–100). Boston, MA: Kluwer Academic Publishers.

[Littman, Sutton, & Singh (2002] Littman, M. L., Sutton, R. S., & Singh, S. (2002). Predictive representations of state. *Advances in Neural Information Processing Systems*, *14*, 1555–1561.

[Llorà & Garrell (2001a] Llorà, X., & Garrell, J. M. (2001a). Inducing partially-defined instances with evolutionary algorithms. *Proceedings of the 18th International Conference on Machine Learning (ICML 2001)*.

[Llorà & Garrell (2001b] Llorà, X., & Garrell, J. M. (2001b). Knowledge independent data mining with fine-grained parallel evolutionary algorithms. *Proceedings of the Third Genetic and Evolutionary Computation Conference (GECCO-2001)*, 461–468.

[Llorà & Goldberg (2002] Llorà, X., & Goldberg, D. E. (2002). *Minimal achievable error in the LED problem* (IlliGAL report 2002015). University of Illinois at Urbana-Champaign: Illinois Genetic Algorithms Laboratory.

[Llorà & Goldberg (2003] Llorà, X., & Goldberg, D. E. (2003). Bounding the effect of noise in multiobjective learning classifier systems. *Evolutionary Computation*, *11*, 279–298.

[Llorà, Goldberg, Traus, & Bernadó (2003] Llorà, X., Goldberg, D. E., Traus, I., & Bernadó, E. (2003). Accuracy, Parsimony, and Generality in Evolutionary Learning Systems via Multiobjective Selection. In Lanzi, P. L., Stolzmann, W., & Wilson, S. W. (Eds.), *Learning classifier system: Fifth international workshop, IWLCS 2002 (LNAI 2661)* (pp. 118–142). Berlin Heidelberg: Springer-Verlag.

[Lobo & Harik (1999] Lobo, F., & Harik, G. (1999). *Extended compact genetic algorithm in C++* (IlliGAL report 99016). University of Illinois at Urbana-Champaign: Illinois Genetic Algorithms Laboratory.

[Mahfoud (1992] Mahfoud, S. W. (1992). Crowding and preselection revisited. *Parallel Problem Solving from Nature*, 27–36.

[Mahfoud (1995] Mahfoud, S. W. (1995). *Niching methods for genetic algorithms*. Doctoral dissertation, University of Illinois at Urbana-Champaign.

[Mitchell, Forrest, & Holland (1991] Mitchell, M., Forrest, S., & Holland, J. H. (1991). The royal road for genetic algorithms: Fitness landscapes and GA performance. In Varela, F. V., & Bourgine, P. (Eds.), *Toward a Practice of Autonomous Systems: First European Conference on Artificial Life* (pp. 245–254). Cambridge, MA: MIT Press.

[Mitchell (1997] Mitchell, T. M. (1997). *Machine learning*. Boston, MA: McGraw-Hill.

[Moore & Atkeson (1993] Moore, A. W., & Atkeson, C. (1993). Prioritized sweeping: Reinforcement learning with less data and less real time. *Machine Learning*, *13*, 103–130.

[Mühlenbein & Paaß (1996] Mühlenbein, H., & Paaß, G. (1996). From recombination of genes to the estimation of distributions I. Binary parameters. *Parallel Problem Solving from Nature - PPSN IV*, 178–187.

[Neal (1993] Neal, R. M. (1993). *Probabilistic inference using Markov chain Monte Carlo methods* (Technical Report CRG-TR-93-1). Dept. of Computer Science, University of Toronto.

[Pashler, Johnston, & Ruthruff (2001] Pashler, H., Johnston, J. C., & Ruthruff, E. (2001). Attention and performance. *Annual Review of Psychology*, *52*, 629–651.

[Pashler (1998] Pashler, H. E. (1998). *The psychology of attention*. Cambridge, MA: MIT Press.

[Pearl (1988] Pearl, J. (1988). *Probabilistic reasoning in intelligent systems: Networks of plausible inference*. San Mateo, CA: Morgan Kaufmann.

[Pelikan (2001] Pelikan, M. (2001). Bayesian optimization algorithm, decision graphs, and Occam's razor. *Proceedings of the Third Genetic and Evolutionary Computation Conference (GECCO-2001)*, 511–518.

[Pelikan (2002] Pelikan, M. (2002). *Bayesian optimization algorithm: From single level to hierarchy*. Doctoral dissertation, University of Illinois at Urbana-Champaign, Urbana, IL. Also IlliGAL Report No. 2002023.

[Pelikan, Goldberg, & Cantu-Paz (1999] Pelikan, M., Goldberg, D. E., & Cantu-Paz, E. (1999). BOA: The Bayesian optimization algorithm. *Proceedings of the Genetic and Evolutionary Computation Conference (GECCO-99)*, 525–532.

[Pelikan, Goldberg, & Lobo (2002] Pelikan, M., Goldberg, D. E., & Lobo, F. (2002). A survey of optimization by building and using probabilistic models. *Computational Optimization and Applications*, *21*(1), 5–20.

[Platt (1998] Platt, J. (1998). Fast training of support vector machines using sequential minimal optimization. In Schlkopf, B., Burges, C., & Smola, A. (Eds.), *Advances in Kernel Methods – Support Vector Learning* (pp. 42–65). Cambridge, MA: MIT Press.

[Poggio & Bizzi (2004] Poggio, T., & Bizzi, E. (2004). Generalization in vision and motor control. *Nature*, *431*, 768–774.

[Potts (2004] Potts, D. (2004). Incremental learning of linear model trees. *Proceedings of the Twenty-First International Conference on Machine Learning (ICML-2004)*, 663–670.

[Quinlan (1993] Quinlan, J. R. (1993). *C4.5: Programs for machine learning.* San Francisco, CA: Morgan Kaufmann.

[Rao & Ballard (1997] Rao, R. P. N., & Ballard, D. H. (1997). Dynamic model of visual recognition predicts neural response properties in the visual cortex. *Neural Computation*, *9*, 721–763.

[Rao & Ballard (1999] Rao, R. P. N., & Ballard, D. H. (1999). Predictive coding in the visual cortex: a functional interpretation of some extra-classical receptive-field effects. *Nature Neuroscience*, *2*(1), 79–87.

[Rechenberg (1973] Rechenberg, I. (1973). *Evolutionsstrategie Optimierung technischer Systeme nach Prinzipien der biologischen Evolution.* Stuttgart-Bad Cannstatt: Friedrich Frommann Verlag.

[Riesenhuber & Poggio (1999] Riesenhuber, M., & Poggio, T. (1999). Hierarchical models of object recognition in cortex. *Nature Neuroscience*, *2*, 1019–1025.

[Riesenhuber & Poggio (2000] Riesenhuber, M., & Poggio, T. (2000). Models of object recognition. *Nature Neuroscience Supplement*, *3*, 1199–1204.

[Rothlauf (2002] Rothlauf, F. (2002). *Representations for genetic and evolutionary algorithms.* Studies in Fuzziness and Soft Computing, 104. Heidelberg: Springer.

[S. Hettich & Merz (1998] S. Hettich, C. L. B., & Merz, C. J. (1998). UCI repository of machine learning databases.

[Sastry & Goldberg (2000] Sastry, K., & Goldberg, D. E. (2000). *On extended compact genetic algorithm* (IlliGAL report 2000026). University of Illinois at Urbana-Champaign: Illinois Genetic Algorithms Laboratory.

[Schwarz (1978] Schwarz, G. (1978). Estimating the dimension of a model. *The Annals of Statistics*, *6*, 461–464.

[Servedio (2001] Servedio, R. A. (2001). *Efficient algorithms in computational learning theory.* Doctoral dissertation, Harvard University, Cambridge, MA.

[Shawe-Taylor & Cristianini (2004] Shawe-Taylor, J., & Cristianini, N. (2004). *Kernel methods for pattern analysis.* Cambridge, UK: Cambridge University Press.

[Simon (1969] Simon, H. A. (1969). *Sciences of the artificial.* Cambridge, MA: MIT Press.

[Simsek & Barto (2004] Simsek, O., & Barto, A. G. (2004). Using relative novelty to identify useful temporal abstractions in reinforcement learning. *Proceedings of the Twenty-First International Conference on Machine Learning (ICML-2004)*, 751–758.

[Singh, Littman, Jong, & Pardoe (2003] Singh, S., Littman, M. L., Jong, N. K., & Pardoe, D. (2003). Learning predictive state representations. *Proceedings of the Twentieth International Conference on Machine Learning (ICML-2003)*, 712–719.

[Stolzmann (1998] Stolzmann, W. (1998). Anticipatory classifier systems. *Genetic Programming 1998: Proceedings of the Third Annual Conference*, 658–664.

[Stone & Bull (2003] Stone, C., & Bull, L. (2003). For real! XCS with continuous-values inputs. *Evolutionary Computation, 11*, 299–336.

[Sutton, Precup, & Singh (1999] Sutton, R., Precup, D., & Singh, S. (1999). Between MDPs and semi-MDPs: A framework for temporal abstraction in reinforcement learning. *Artificial Intelligence, 112*, 181–211.

[Sutton (1990] Sutton, R. S. (1990). Integrated architectures for learning, planning, and reacting based on approximating dynamic programming. *Proceedings of the Seventh International Conference on Machine Learning, 7*, 216–224.

[Sutton (1991] Sutton, R. S. (1991). Reinforcement learning architectures for animats. *From Animals to Animats: Proceedings of the First International Conference on Simulation of Adaptive Behavior*, 288–296.

[Sutton & Barto (1998] Sutton, R. S., & Barto, A. G. (1998). *Reinforcement learning: An introduction*. Cambridge, MA: MIT Press.

[Sutton & Singh (2004] Sutton, R. S., & Singh, S. (2004). Proceedings for the ICML'04 workshop on predictive representations of world knowledge.

[Taylor (2002] Taylor, J. G. (2002). From matter to mind. *Journal of Consciousness Studies, 9*, 3–22.

[Thierens & Goldberg (1993] Thierens, D., & Goldberg, D. E. (1993). Mixing in genetic algorithms. *Proceedings of the Fifth International Conference on Genetic Algorithms*, 38–45.

[Thierens, Goldberg, & Pereira (1998] Thierens, D., Goldberg, D. E., & Pereira, A. G. (1998). Domino convergence, drift, and the temporal-salience structure of problems. In *Proceedings of the 1998 IEEE World Congress on Computational Intelligence* (pp. 535–540). New York, NY: IEEE Press.

[Tolman (1932] Tolman, E. C. (1932). *Purposive behavior in animals and men*. New York: Appleton.

[Valiant (1984] Valiant, L. (1984). A theory of the learnable. *Communications of the ACM, 27*, 1134–1142.

[Venturini (1994] Venturini, G. (1994). Adaptation in dynamic environments through a minimal probability of exploration. *From Animals to Animats 3: Proceedings of the Third International Conference on Simulation of Adaptive Behavior*, 371–381.

[Watkins (1989] Watkins, C. J. C. H. (1989). *Learning from delayed rewards*. Doctoral dissertation, King's College, Cambridge, UK.

[Whitehead & Ballard (1991] Whitehead, S. D., & Ballard, D. H. (1991). Learning to perceive and act by trial and error. *Machine Learning, 7*(1), 45–83.

[Wilson (1985] Wilson, S. W. (1985). Knowledge growth in an artificial animal. *Proceedings of an International Conference on Genetic Algorithms and Their Applications*, 16–23.

[Wilson (1987a] Wilson, S. W. (1987a). Classifier systems and the animat problem. *Machine Learning*, *2*, 199–228.

[Wilson (1987b] Wilson, S. W. (1987b). The genetic algorithm and simulated evolution. In Langton, C. G. (Ed.), *Artificial Life*, Volume VI (pp. 157–166). Addison-Wesley.

[Wilson (1991] Wilson, S. W. (1991). The animat path to AI. *From Animals to Animats: Proceedings of the First International Conference on Simulation of Adaptive Behavior*, 15–21.

[Wilson (1994] Wilson, S. W. (1994). ZCS: A zeroth level classifier system. *Evolutionary Computation*, *2*, 1–18.

[Wilson (1995] Wilson, S. W. (1995). Classifier fitness based on accuracy. *Evolutionary Computation*, *3*(2), 149–175.

[Wilson (1998] Wilson, S. W. (1998). Generalization in the XCS classifier system. *Genetic Programming 1998: Proceedings of the Third Annual Conference*, 665–674.

[Wilson (2000] Wilson, S. W. (2000). Get real! XCS with continuous-valued inputs. In Lanzi, P. L., Stolzmann, W., & Wilson, S. W. (Eds.), *Learning classifier systems: From foundations to applications (LNAI 1813)* (pp. 209–219). Berlin Heidelberg: Springer-Verlag.

[Wilson (2001a] Wilson, S. W. (2001a). Function approximation with a classifier system. *Proceedings of the Third Genetic and Evolutionary Computation Conference (GECCO-2001)*, 974–981.

[Wilson (2001b] Wilson, S. W. (2001b). Mining oblique data with XCS. In Lanzi, P. L., Stolzmann, W., & Wilson, S. W. (Eds.), *Advances in learning classifier systems: Third international workshop, IWLCS 2000 (LNAI 1996)* (pp. 158–174). Berlin Heidelberg: Springer-Verlag.

[Wilson (2002] Wilson, S. W. (2002). Classifiers that approximate functions. *Natural Computing*, *1*, 211–234.

[Wilson (2004] Wilson, S. W. (2004). Classifier systems for continuous payoff environments. *Proceedings of the Sixth Genetic and Evolutionary Computation Conference (GECCO-2004): Part II*, 824–835.

[Wilson & Goldberg (1989] Wilson, S. W., & Goldberg, D. E. (1989). A critical review of classifier systems. *Proceedings of the Third International Conference on Genetic Algorithms*, 244–255.

[Witten & Frank (2000] Witten, I. H., & Frank, E. (2000). *Data mining. Practical machine learning tools and techniques with java implementations*. San Francisco, CA: Morgan Kaufmann.

[Wolpert (1996a] Wolpert, D. H. (1996a). The existence of a priori distinctions between learning algorithms. *Neural Computation*, *8*(7), 1391–1420.

[Wolpert (1996b] Wolpert, D. H. (1996b). The lack of a priori distinctions between learning algorithms. *Neural Computation*, *8*(7), 1341–1390.

[Wolpert & Macready (1995] Wolpert, D. H., & Macready, W. G. (1995). *No free lunch theorems for search* (Tech. Rep. No. SFI-TR-95-02-010). Santa Fe, NM: Santa Fe Institute.

[Wolpert & Macready (1997] Wolpert, D. H., & Macready, W. G. (1997). No free lunch theorems for optimization. *IEEE Transactions on Evolutionary Computation*, *1*(1), 67–82.

Index

ε-greedy action selection   20, 34, 228,
    229, 235

ACS2   201, 203, 209
ALCS   208, 209
animat   1
anticipation   7, 208, 215, 216
anticipatory behavior   X, 5, 208,
    214–216
attention   208, 212, 214–216

Bayesian
  dirichlet metric   135, 136
  information criterion   135, 136
  model   28, 124, 144, 145
  networks (BNs)   134–137, 139, 221
  optimization algorithm (BOA)   4,
      28, 124, 131, 132, 134–136, 210
  recombination in XCS   152
  search in XCS   124, 137, 141–145,
      147–149, 152, 153, 221
behavioral policy   21, 29, 32, 34, 38, 49,
    53, 181, 189, 193, 195, 219, 221
blocks world problems   186, 192, 193
building blocks (BBs)   23
  hard problems   4, 124, 221, 224
  identification   130–135, 137
  in population   25, 26
  of cognition   215, 216, 225
  of problem   9, 11, 12, 14, 23, 126, 127,
      137, 139, 248
  of solution   4, 44, 48, 213, 220, 221,
      245
    loosely coded   128, 144

  tightly coded   127–131
processing   27, 29, 94, 123, 124,
    126–130, 144, 248
recombination   140, 142–144, 210
structure   27, 28, 94, 122, 137, 138,
    146

classification problems   5, 6, 9, 10,
    13–16, 28, 29, 39, 49, 54, 58, 61, 74,
    94, 122, 123, 219
  Boolean functions   14, 15, 66, 147,
      148, 221
    XCS performance   148–156
  definition   227
  hierarchical   123–127, 144, 221
  real-valued   15
  simple example   58–60
cognition   VII, 1, 2, 207, 215, 216
cognitive sequence learning (COSEL)
    214
cognitive structures   207, 211
  learning   7, 223, 225
cognitive system   7, 211, 215, 223
  aspects   207
  CS1   VII, 1, 206, 225
  LCSs   5, 206
  learning   5, 207, 214, 216, 217, 225
covering   35, 53, 56, 110, 111, 157, 158,
    161, 163, 165, 170–173, 179, 231,
    233, 234
  alg.description   234
  bound   91–95, 98, 100, 101, 104, 118,
      119, 121, 142, 160, 205
  definition   53

crossover   22, 25–28, 36, 41, 43, 44, 46,
    49, 50, 57, 86, 89, 107, 109, 112,
    113, 121, 123, 142, 154, 204
  example   44, 45
  informed   128, 129, 131, 142, 144
  model-based   28, 122–124, 129, 130,
    141, 142, 144, 145, 149, 204, 209
  one- or two-point   27, 128–131, 144,
    204
    definition   27
    example   36
  real-valued   157, 160, 170–173, 179,
    205, 206
  simple   27, 138, 140, 141, 144, 151,
    205, 209
  uniform   27, 35, 108, 125, 128–130,
    141, 142, 144, 148, 149, 152, 154,
    170–173, 186, 204
    alg.description   239
    definition   27

datamining
  problems   6, 93, 122, 157, 160, 163,
    164, 181
    definition   15
    XCS performance   165, 167, 168,
    222
  system   157
  types   32
deletion pressure   67, 70, 73, 74, 78, 89
diversification pressure   50
DYNA   21

evolution
  Darwinian   1, 23, 29
  learning   IX, 31, 35, 108, 131, 149,
    151, 154, 171, 186, 197, 200, 219,
    220, 225
  pressure   22, 34, 48, 65–68, 70–78, 80,
    88, 91, 92, 118, 159, 204, 220, 221
  representation   32, 211
  search   5, 22, 43, 152, 179, 204, 221,
    225
  solution   1, 141, 160
  strategies   21, 35, 158
  structure   9, 35, 176, 207, 209, 214
exponentially scaled problem   12
extended compact GA (ECGA)   4, 28,
    124, 131–133, 210

marginal product model   124, 132,
    133
  search in XCS   133, 137, 141–145

fitness guidance   11, 23, 38, 39, 48, 66,
    71, 78, 88, 91, 103, 106, 107, 118,
    124–126, 147–152, 154, 162, 243,
    244, 246, 247
fitness pressure   6, 40, 65, 67, 71, 73, 74,
    78, 80–82, 84–89, 91, 95, 125, 147,
    148, 154, 160, 220, 221
function approximation problems   6,
    122, 169–172, 179, 222
  XCS performance   169, 172–178

generalization pressure   40, 57, 60, 63,
    65, 66, 85, 154, 220
  in ZCS   200
  semantic   63, 67, 68, 72–74, 88, 200,
    201, 221
  syntactic   200, 221

hierarchy
  default   2, 162
  learning   207, 208, 212–216, 219, 223,
    225
  problems   9, 123–130, 141–145, 151,
    221, 224, 248
  representation   5, 208, 214–216
  structures   X, 7, 127, 212, 214, 215,
    217
hyperellipsoidal conditions   169–171,
    205
  definition   229
  general   171, 172, 205
    definition   229
    XCS performance   177, 178
  XCS performance   176–178
hyperrectangular condition   205
  in binary problems   42, 52, 64
  in real-valued problems   158, 159,
    169, 177
    definition   229
    XCS performance   173, 174
hyperspheroidal conditions   169, 170,
    205
  definition   229
  XCS performance   173–176, 205

Markov decision process

MDP   9, 10, 16–18, 21, 29, 186, 227
  partially observable (POMDP)   10,
    16–18
maze problems   3, 18, 186–192
  Maze 1   16, 17, 20, 33, 36, 38, 39, 45,
    61, 62, 183, 187
  Maze 2   17
  Maze 5   187–191
  Maze 6   187–189, 192
  Woods14   187, 190, 191
minimum description length (MDL)
    132, 133, 137
mutation   3, 22, 25–29, 35, 36, 41, 42,
    47–50, 57, 60, 65, 69, 72, 94, 95, 97,
    104, 105, 109, 112, 113, 122, 123,
    128–131, 141, 144, 204, 205, 220,
    221, 246
  example   42, 43
  free   69, 70, 74, 75, 78–80, 108, 153,
    230
    alg.description   240
    definition   57
  generalization   42, 59, 63, 72, 201
  generalizing   201
  niche   69, 70, 74, 75, 79, 80, 107, 108,
    230
    definition   57
  pressure   67, 69, 72–75, 84, 89, 153
  rate   74, 75, 77–80, 82, 83, 85, 88, 98,
    100, 104, 107, 118, 121, 129, 130,
    141, 142, 144, 148, 154–156, 204
  real-valued   157–160, 163, 170–172,
    179, 205, 206
  specialization   41–43, 72, 75, 126,
    188, 201, 204
  type   70, 72, 75, 78, 80, 128, 158

needle in the haystack problem   11–13
neural networks   5, 48, 209, 211, 224
  RL   5, 7, 182, 194, 195, 222
neural processing   212, 215
neural structures   211, 213
niche
  deletion   203
  distribution   14, 108–115, 120
  occurrence   113, 117, 119, 120, 157,
    159, 181, 200, 202, 221, 235
  of problem   25, 70, 71, 92, 93,
    104, 109, 113, 114, 119, 130, 132,

    137, 139, 140, 146, 152, 160, 181,
    198–202, 204, 221, 247
  definition   109
  overlapping   70, 113–115, 130
  representative   109, 115
  reproduction   3, 51, 57, 63, 66, 111,
    131, 202, 203, 220, 221
  selection   202, 203
  support   70, 203, 248
  support bound   108–111, 116, 117,
    120, 122, 162
  sustenance   26, 48, 182, 186, 189, 199,
    202, 203
niching   26, 29, 46–48, 58, 63, 92, 108,
    147, 199, 202, 203, 220, 241
  crowding   26, 29
  in ZCS   203
  sharing   26, 29

observable operator models (OOMs)
    208
one-max problem   11–13, 125, 153
optimization problems   5, 9–14, 22, 23,
    28

PAC learning   4, 92, 117–122, 219
predictive state representation (PSR)
    208
probabilistic logic sampling (PLS)   135
problem difficulty
  k_d 106, 244, 246, 248
    definition   227
  minimal order of k_m 95, 96, 100,
    101, 103, 119, 121, 122, 124, 146,
    152, 153, 161, 221, 244, 246–248
    definition   94, 227
proportionate selection
  in GAs   24, 25
  in LCSs   35, 197, 198, 202, 203
  in XCS   56, 79–84, 88, 89, 148, 154,
    155, 165, 166
    alg.description   238, 241

Q-learning   3, 9, 18–21, 29, 35, 181, 182,
    220
  approximation   183, 184
  in LCS1   34, 35
  in XCS   54, 58, 61–63, 89, 184, 185,
    221, 235
  tabular   5, 182, 185, 195, 224

reinforcement learning (RL)   IX, 3,
      5, 9, 10, 18, 29, 31, 32, 47–49, 54,
      58, 61, 63, 179, 182, 193, 207, 213,
      220–222, 235
   approximation   5, 182, 183
   component in LCSs   32–35, 40, 41,
      47, 49, 221
   model-based   21
   model-free   18–20
   tabular   182, 183, 194
reproductive opportunity   24, 59, 60,
      63, 67, 84, 104, 148, 156, 157, 162,
      182, 189, 190, 198, 200
   bound   91, 92, 98–104, 108, 119,
      121–124, 129, 142, 149, 160, 194,
      205
RL problems   5, 6, 9, 10, 15, 16, 29, 45,
      49, 53, 54, 58, 61, 62, 66, 93, 118,
      181, 195, 198, 219, 222, 227
   multistep   5, 6, 16, 17, 122, 181, 182,
      219, 221, 222, 232, 244
   single-step   16, 18, 74, 122, 144, 232,
      235
royal-road function   12
royal-road function   11–13

SARSA   2
schema
   bound   91, 94–98, 100, 101, 103, 104,
      109, 119, 122–124, 160, 205
   definition   94
   growth   48, 91, 95, 160

order, definition   94
   representative   91, 95, 96, 99, 100,
      123
      definition   94
   supply   48, 91, 94, 95, 97, 123, 157,
      160
schema theory   4, 128, 204
schemata   3, 4, 97, 99, 116, 124, 246
selection pressure   24, 25, 28, 80, 83, 84,
      86, 87, 181
set pressure   66–68, 72–74, 82, 84, 88,
      89, 159, 160, 201
specialization pressure   85, 140, 153
subsumption pressure   67, 70, 71, 73,
      74, 89, 160, 221

TD(λ)   2, 18, 19, 21
time bound   92, 104–108, 119, 121–123
tournament replacement   203
tournament selection
   in GAs   24–26
   in LCSs   197, 201, 202
   in XCS   6, 79–81, 84–89, 91, 98, 109,
      140, 147, 154, 155, 163, 165, 166,
      186, 194, 221
      alg.description   239
trap problem   11–13, 23, 94, 134

uniformly scaled problem   12

ZCS   2, 3, 39, 40, 198–201, 203